Biocoordination Chemistry

David E. Fenton

Department of Chemistry,
University of Sheffield

OXFORD
UNIVERSITY PRESS

Contents

1 'Life is inorganic too'

Although traditionally biology, and hence life, has been regarded as organic, life is inorganic too. While you are reading this you are breathing, taking in dioxygen from the atmosphere and using it for the essential process of respiration through its interaction with the iron atoms present in haemoglobin and myoglobin. In order to read this you are continually sending messages throughout your nervous system by using the changes in electrical flux caused by the movement of sodium and potassium ions across cell membranes. It has been said that 'Life has evolved from inorganic materials (generating organic chemistry as it went) and in that evolution has incorporated every facet of inorganic chemistry that was profitable to it.' Every facet of inorganic chemistry but not every element; at present we are aware of 13 metals which are essential for plants or animals. Four of these, sodium, potassium, magnesium, and calcium, are present in large quantities and are known as the *bulk metals*; the remaining nine which are present in small quantities are the d-block elements vanadium, chromium, molybdenum, manganese, iron, cobalt, nickel, copper, and zinc and are known as the *trace metals*. It is this latter group that we are concerned with in biocoordination chemistry.

1.1 Nonmetallic elements

The nonmetallic elements provide, through water, the medium of life. Carbon, nitrogen, hydrogen, and oxygen provide the fundamental organic components of life by making up molecular building blocks such as amino acids, which themselves are the building blocks for proteins, sugars, fatty acids, purines, pyrimidines, and nucleotides (Fig. 1.1). Phosphorus also plays

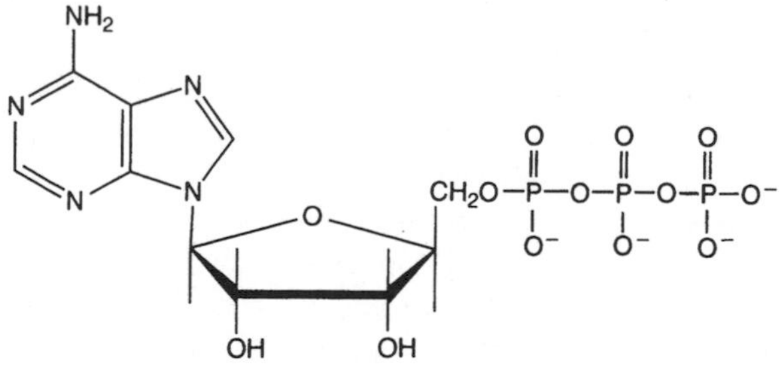

Fig. 1.1. Nucleotides contain the following covalently bound sequence, base–carbohydrate–phosphate; ATP serves as a representative example.

an important role in adenosine triphosphate (ATP) which is central to the energetics of the cell. It is also present in the double helix of deoxyribonucleic acid (DNA), a polymer of nucleotides, which incorporates the genetic blueprint of all organisms except for RNA viruses. The nonmetals also provide counter-ions such as Cl^-, SO_4^{2-}, CO_3^{2-}, and PO_4^{3-} ; mineralized tissues such as bones and teeth are derived from calcium phosphates, and shells from calcium carbonate, together with a complicated organic matrix of polysaccharides, lipids, and proteins.

1.2 Metals in biological systems

The bulk metals form 1–2% of human body-weight and a 75kg person has approximately 170g potassium, 100g sodium, 1100g calcium, and 25g magnesium present. In contrast the trace elements represent less than 0.01% of human body-weight and even of iron, the most widely used, we need only of the order of 4–5g.

Two major periods of activity in trace metal research have occurred during the twentieth century. Until 1956, serendipity aided the discovery of the importance of copper, zinc, cobalt, manganese, and molybdenum in animals; since 1956, work based on the experimental induction of trace element deficiencies, inspired by Klaus Schwarz, indicated that other metals such as chromium, tin, vanadium, nickel, cadmium, and lithium should perhaps be added to the list. One problem is that living organisms are adept at accumulating metals from media of extreme dilution, from airborne dust and by leaching from feeding vessels. Accumulation in itself does not establish that there is a physiological need for a metal. Marine organisms are known with high titanium, vanadium, chromium, niobium, and thallium content and mussels are currently used to monitor the purity of rivers and estuarine waters. There is still conjecture concerning the possible essentiality of cadmium and lithium, and the mammalian metabolism of cadmium may be as a detoxificant through the induction of *metallothionen*. The *phytochelatins* are the plant world's equivalent of metallothionen and bind strongly to cadmium and other heavy metals. (The biomolecules in italic face will be described later in the text.)

This is a living subject area and so it is interesting to speculate on whether, or not, further essential elements will be discovered.

1.3 Trace metals

The trace metals may be divided into two subgroups. Iron, copper, and zinc form one group; the remaining six metals are termed the ultra-trace elements, being present in exceedingly small concentrations.

Iron, copper, and zinc, the three most abundant trace elements, are universally essential. Iron and copper can readily change their oxidation states and so act in electron-transfer systems; these can produce oxidized substrates which participate in a variety of metabolic cycles. Examples include the Fe–S proteins *rubredoxin* and *ferredoxin*, the haem protein *cytochrome* c, and the blue copper proteins *azurin* and *plastocyanin*. Iron and copper are also involved in dioxygen storage and carriage. The former is present in *haemoglobin*, *myoglobin*, and *haemerythrin* and the latter in *haemocyanin*. There are iron storage (*ferritin*) and carrier proteins (*transferrin*, *lactoferrin*, and *siderophores*) and a copper-transfer protein (*ceruloplasmin*). Zinc serves as a superacid centre in metalloenzymes

Metal	g/75kg
Na	70–120
K	160–200
Ca	1100
Mg	25
V	15×10^{-3}
Cr	2×10^{-3}
Mn	1
Fe	4–5
Co	1.2×10^{-3}
Cu	$80–120 \times 10^{-3}$
Zn	2–3
Mo	10×10^{-3}

Metals in man

1.17

1.18 R = p--CF$_3$C$_6$H$_4^-$

1.19 Pyr = pyridine

1.20

1.21

1.22

not yet been achieved. This may be associated with the concept of the entatic state, namely that the active metal site of an enzyme is in a geometry approaching that of the transition state of the appropriate reaction and as such is uniquely fitted for catalytic action. The iron–sulphur clusters (**1.17**, **1.18**) synthesized to mimic the Fe–S proteins (**1.9**, **1.11**) are examples of corroborative models. The dinuclear copper(II) complex (**1.19**) is representative of the many speculative models which were advanced for the dinuclear copper(II) site in oxyhaemocyanin as predicted from cumulative spectroscopic studies (**1.20**). It was, however, the peroxide binding mode in another speculative model complex (**1.21**) that was found when the crystal structure of oxyhaemocyanin was solved (**1.22**).

One necessary statement concerning the use of models is that caution must be exercised. It is essential to remember that the model is a small molecule and so is not in any way constrained by the immediacy of a protein environment and the accompanying subtle roles that the folded protein chains may play in governing the nature of the biosite. Therefore while a model may accurately reproduce the physicochemical properties of a site, it may not be able to respond in a truly functional manner. Perhaps this is best summarized by the econometrician Thiel who commented that 'Models are to be *used*, not believed'.

promoting the hydrolysis or cleavage of a variety of chemical bonds; representative examples are *carboxypeptidase* (cleaves terminal carboxy groups from peptide chains), *carbonic anhydrase* (regulates the hydration of carbon dioxide), and *alcohol dehydrogenase* (converts alcohol to acetaldehyde). As well as being involved in functional roles, zinc can also act in a structural role in which it controls and stabilizes the enzyme structure by helping to position the enzyme for action at a distant site. This is exemplified in the bimetallic enzyme *superoxide dismutase* which breaks down toxic superoxide to give dioxygen and peroxide, the latter of which is converted into dioxygen and water by the iron-containing metalloenzyme *catalase*.

Bovine erythrocyte superoxide dismutase (BESOD), *haemocyanin*, and *haemerythrin* may be used to introduce bimetallobiosites wherein there is a bimetallic centre in the biological molecule. BESOD has a heterobimetallic centre consisting of one zinc and one copper atom; haemocyanin and haemerythrin both have homobimetallic centres bearing two copper and two iron atoms respectively.

1.4 The ultratrace metals

Of the so-called ultratrace metals only five (Mn, Mo, Co, Ni, and V) have been identified as forming metalloenzymes. Manganese is involved in several important enzymes—*mitochondrial superoxide dismutase*, *inorganic phosphatase*, *glycosyl transferase*, and in *photosynthesis system II*. Cobalt is present in *vitamin B$_{12}$* and its coenzyme and provides a rare example of a naturally occurring organometallic compound. Nickel is now known to function in metalloenzymes such as *urease* and several *hydrogenases* and to act in the unusual nickel(III) oxidation state. Both molybdenum and vanadium are found in *nitrogenases*, where they are present in clusters containing also iron and sulphur; vanadium is also active in *haloperoxidases*. The remaining metal, chromium, has persistently been reported as being essential in glucose metabolism in higher mammals but there has not yet been any definite verification of this essentiality.

1.5 The roles of metal ions in biological systems

A useful classification of metallobiomolecules is given in Fig.1.2. The potential roles of metal ions in biological systems can be described as *structural* and *functional*. In the former the metal ion helps to stabilize the protein structure and in the latter the metal ion is involved in the reactivity of the biosite. Proteins are polymers of amino acids joined by amide bonds. Members of this class of compound are also referred to as peptides or polypeptides. In this book proteins are regarded as naturally occurring polypeptides and peptides are considered as smaller molecules with defined degrees of association—dipeptides (two amino acids), tripeptides (three amino acids) and so on. Enzymes act as catalysts enabling reactions to occur under physiological conditions which would otherwise be unacceptably slow. All enzymes are proteins but all proteins are not necessarily enzymes; metalloproteins and metalloenzymes contain at least one metal ion.

amide bond

1.8 The entatic state

Structural studies on metalloenzymes have shown that coordination geometries around the metal ions can be distorted. It has been proposed that this may be related to the catalytic efficiency of the enzyme and that the protein can fold to generate special stereochemistries which help the metal to adopt a geometry closer to that in the transition state of the reaction being catalysed than to a more normal regular state geometry. There is 'a presetting of the geometry such that there is a catalytically poised state intrinsic to the active site' (Fig. 1.3). This has been termed the entatic state (*entasis* - Gk, stretched, under tension) and examples include the copper atoms in azurin and plastocyanin, and the zinc atoms at the active sites of carbonic anhydrase and carboxypeptidase.

1.9 Why small molecule models for biosites?

It is not always possible to have such a precise awareness of the nature of the metallobiosite. Often it is necessary to build up a picture of the site based on a cumulation of spectroscopic evidence and this has led to the study of small molecule models for the biosites. Models have been classified as being of three types: *corroborative*, where the structure of the biosite is known and the model is built to study properties of the site *in vitro* and to determine if the properties of the metalloprotein are dominated by the first coordination sphere of the metal; *speculative,* where the structure is not known but is anticipated from cumulative spectroscopic studies and so the model is used to parallel these properties in order to produce a predictive comparison; and *functional*, where the actual functioning of the site is reproduced. This latter aspect is deemed to be the most difficult and it is still true to say that it has

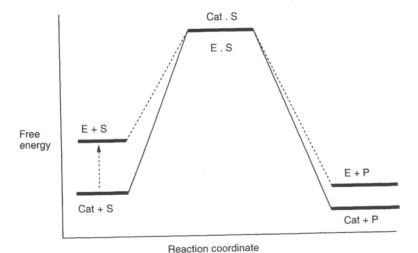

Fig.1.3. Representation of the way in which the reaction profile differs between an enzyme catalyst (E) and a homogeneous catalyst (Cat) [S = substrate and P = product] (after Williams).

2 Metal management

After the uptake of a metal into an organism from the environment, the metal must be transported to the site where it is to be incorporated for use. Devices for storing excess metals until there is a requirement for use are also necessary. The storage and transport of iron is now quite well understood, but for other metals there is a paucity of detailed information.

2.1 Iron storage

Ferritin

The storage of iron in a non-toxic form is a necessary feature of metabolism as iron is required for the production of functional molecules such as Fe–S and haem proteins. Free, unbound iron is extremely toxic as it can catalyse free radical formation and so cause significant cellular damage. In man the iron storage system must be able to respond to 'supply and demand' as the amount of iron in the diet is variable. If iron is in excess then it must be stored but if iron is lost from the body then the store must be mobilized and replenish this loss.

The storage reservoir for iron is ferritin which is widely distributed in various organs of mammals, particularly in the liver, spleen, and bone marrow, and is also present in plants and bacteria. Ferritin consists of a large hollow protein encasing a mineral core of hydrated iron(III) oxide, a 'particle of rust'. A crude analogy is an iron tablet of the type prescribed for iron deficiency which is 'sugar coated' to make it palatable and to allow slow release of the iron. The diameter of the core is some 80Å and contains up to 4500 Fe(III) atoms. The composition of the microcrystalline core is $(FeOOH)_8(FeOH_2PO_4)$ and the X-ray diffraction of the core is similar to that of ferrihydrate, $5Fe_2O_3 \cdot 9H_2O$. An EXAFS study suggests that the structure is based on a close-packed array of oxide and hydroxide anions with iron atoms occupying octahedral interstices. The phosphate appears to act in a terminal role perhaps by covering the (FeOOH) particles and anchoring them to the protein shell (Fig. 2.1).

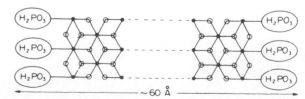

Fig. 2.1. One way of using phosphates to terminate the two-dimensional Fe–O sheets in the ferritin core (reproduced with permission from the American Chemical Society).

Extended X-ray absorption fine-structure (EXAFS) spectroscopy has proved to be a very useful complementary technique to X-ray crystallography for the elucidation of the structures of metallobiosites. This is because it does not require a crystalline sample. Synchrotron radiation is used to excite a core electron of the metal at the biosite to the ionization continuum. At the metal K absorption edge one observes a series of peaks, due to the transition of the 1s electron into molecular orbitals having mainly transition metal character, superimposed on a steeply rising absorption due to transition of the 1s electron into continuum levels. The shape of the edge and the positioning of the edge features can give information about the coordination environment of the metal. Mathematical analysis and comparisons with the spectra of complexes of known coordination environment yields information about the number and nature of the donor atoms at the site. The technique differentiates between first- and second-row donors but not between first-row donors, and it does not give angular information. The accuracy is less than that of X-ray crystallography—bond lengths can be determined to ± 0.02Å.

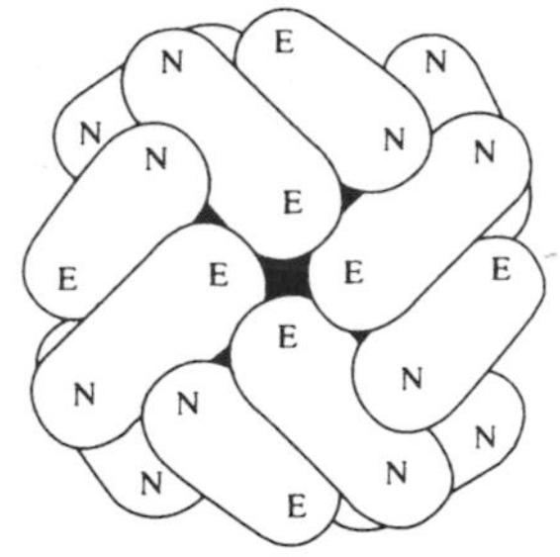

Fig. 2.2. Schematic drawing of the subunits in horse spleen apoferritin viewed down a fourfold axis (reproduced with permission from Academic Press).

The protein sheath of the core free apoferritin has been shown by Pauline Harrison and her research group, using X-ray crystallography, to consist of 24 subunits, each of which is a polypeptide chain coiled in a lozenge-like shape (Fig. 2.2). There are two different polypeptide chains termed H– and L–; the former predominates in heart and brain ferritins, with liver and spleen ferritins being L-rich.

The mechanisms by which the iron core is formed and the iron leaves ferritin are still not well understood. What is remarkable is that the core can only be formed from aqueous iron(II) and that oxidation to iron(III) follows incorporation. It has been proposed that the iron gets into the core through channels in the protein and then is transferred into the cavity to form first diiron–oxo dimers and then aggregates and clusters are formed via a progression of oligomers related to iron hydrolysates. An iron(III)– tyrosinate complex has been identified in ferritin by UV and resonance Raman spectroscopy and the proposal has been made that this may be a transient precursor to polynuclear cluster formation.

Biomineralization is the process by which substances such as shells, skeletons, and teeth are built. The fascination of ferritin is that the protein provides a framework within which the biomineralization process occurs as the iron is deposited giving a relatively inert material which provides a storage reservoir.

Haemosiderin

Iron is also stored in a number of mammalian organs as haemosiderin which contains larger amounts of 'iron hydroxide' than ferritin. Its constitution is variable and it is thought that haemosiderin may be a degradation product of ferritin. Haemosiderin with iron cores similar to the ferrihydrite cores of ferritin has been identified in haemosiderins isolated from alcoholic livers and from livers with naturally occurring, or induced, iron overloads.

Synthetic iron–oxo aggregates

By controlling the extent of hydrolysis of iron(III) a variety of oxo- and hydroxobridged oligomers related to the polynuclear iron–oxo units found in biological systems have been prepared. Investigations by Stephen Lippard and his research group into the 'bioinorganic chemistry of rust' have assisted in our attempts to understand the mechanisms of the assembly of high nuclearity iron–oxo clusters. Using $[Et_4N]_2[Fe_2OCl_6]$ and $[Et_4N][FeCl_4]$ as building blocks, tri- and tetranuclear aggregates were synthesized and a much higher aggregate, $[Fe_{11}O_6(OH)_6(O_2CPh)_{15}]$, was formed when an acetonitrile solution of $[Et_4N]_2[Fe_2OCl_6]$ and sodium benzoate was slowly crystallized. The elegant simplicity of this remarkable complex is explained in Fig. 2.3. The eleven iron atoms, which are found at the vertices of a distorted, pentacapped trigonal prism with threefold symmetry, are stacked 1:3:3:3:1 along the three-fold axis. Six triply bridging oxygen atoms connect the iron atoms leading to eighteen short Fe–O bonds and six pyramidal triply bridging oxygen atoms further link the iron atoms in layers, giving rise to eighteen Fe–OH bonds. The structure is completed by fifteen bridging benzoate groups giving a further thirty Fe–O bonds and each of the eleven iron atoms has a distorted octahedral coordination geometry. The extent to which the synthetic core structures resemble that in ferritin is not yet established.

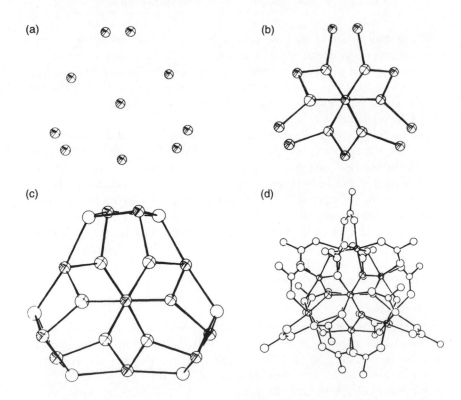

Fig. 2.3. The stepwise assembly of the structure of $[Fe_{11}O_6(OH)_6(O_2CPh)_{15}]$ (a) iron(III) atoms viewed down the threefold axis; (b) addition of μ_3-oxo atoms; (c) addition of the μ_3-hydroxo atoms; (d) full structure (reproduced with permission from the Royal Society of Chemistry).

2.2 Iron transport

Under physiological conditions the thermodynamically stable form of iron is iron(III). Hydrated and uncomplexed iron(III) ions are readily hydrolysed to form insoluble iron hydroxide aggregates and so nature has evolved sophisticated transport systems to complex iron and keep it soluble. To obtain the iron that they need to survive, microbial cells synthesize low molecular weight phenolate and hydroxamate compounds called siderophores; in higher animals the transferrins fulfil this function.

Transferrins

Serum transferrin (the iron transport protein), lactoferrin (present in milk), and ovotransferrin, or conalbumin, (found in egg white) are monomeric glycoproteins having $M_r \sim 80,000$ which bind two iron atoms tightly, but reversibly, as iron(III) with binding constants in excess of $10^{20}M^{-1}$. An interesting feature is that there is a requirement for anion binding to the metal with carbonate being the highest affinity anion and there is a charge transfer ascribable to phenolate ligands to iron which accounts for their salmon pink colour.

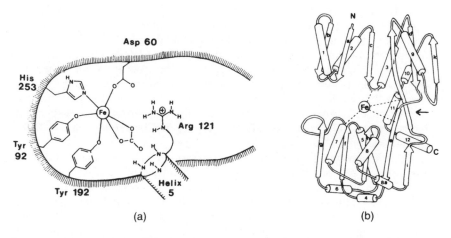

(a)   (b)

Fig. 2.4. (a) Schematic drawing of the iron and anion binding sites in human lactoferrin; (b) polypeptide chain folding in each lobe of human lactoferrin—the possible hinge point is arrowed (reproduced with permission from *Pure and Applied Chemistry*).

The iron site in human lactoferrin has been defined by X-ray crystallography (Edward Baker and co-workers). It consists of an iron atom coordinated to four residues from the protein, two tyrosines, one aspartate, and one histidine. A bidentate carbonate anion occupies the remaining two sites of the distorted octahedral environment. The anion sits in a pocket between the iron atom and two positively charged protein groups, an arginine side chain, and a helix N-terminal (Fig. 2.4a). The iron site in rabbit serum transferrin has been shown to be closely similar to the above. There is a preference for carbonate as the synergistic anion but other carboxylates can also bind at this site.

The iron is buried in a deep cleft between two protein domains and there are two polypeptide strands running behind it. A flexing of these strands leading to an open conformation is envisaged as providing a mechanism for releasing the iron. It is thought that in order to incorporate the iron it binds first to the domain at which the carbonate is bound and then the second domain folds on top (Fig. 2.4b). The role of the carbonate is a puzzle but it is possible that it helps in the organization of the cleft by binding there in advance of the iron.

When microbes invade living tissue, a battle ensues between host and invader for the available iron. If the host iron levels are higher than normal, this causes the host to have an exceptional susceptibility to bacteria and fungi and so the function of the transferrin is to protect the host by denying iron to the invaders. For example, lactoferrin is a potent antibacterial transferrin and acts to protect breast-fed infants against certain infectious disease. One physiological function of fever might be to suppress the synthesis of bacteriological siderophores as it has been observed that siderophore synthesis is sharply reduced at temperatures above 37°C.

Siderophores

Aerobic bacteria use siderophores to make soluble and transport iron as iron(III). These are low molecular weight compounds with remarkably high affinities for the metal ($K_F \sim 10^{30-50} M^{-1}$). They fall into two general categories, representative examples of which are depicted in Fig. 2.5, one having hydroxamate ligands (ferrichromes, ferrioxamines) (**2.1**) and the other having catecholate ligands (enterobactin) (**2.2**) and the iron atom is octahedrally bound in either linear or cyclic molecules. There are also variations on these binding themes with mixed ligand sites, for example mycobactin has two hydroxamate ligands and one phenolate ligand, and aerobactin contains hydroxamate and citrate sites. Each category can be seen to have a 'hard' acid to 'hard' base pairing.

Tris-chelated octahedral complexes can provide optical isomers and so the siderophores can be chiral. For example, molecular models of ferrioxamine B (R = H, $n = 5$, R' = CH$_3$) show that there are five possible isomers for each of the absolute configurations Δ- and Λ- available to the complex. The Λ-form follows the movement of a left-handed propeller (counter-clockwise) when viewed along the C_3 axis of the octahedron and for the Δ-form the movement is right-handed (clockwise). The isomers may be drawn according to the following convention that in each chelate ring if the C atom is below the N atom then it is called C, and rings 2 and 3 are called *cis* and *trans* depending upon which has the same or opposite orientation to the C_3 axis as ring 1. The five Λ-isomers are shown in Fig. 2.6.

The ferrichromes, ferrioxamines, and aerobactin are left-handed with Λ-*cis* configurations; enterobactin and rhodotorulic acid are right-handed with Δ-*cis* configurations. Iron(III) complexes, having the metal present in high-spin d^5 electronic configuration, have no crystal field stabilization energy (CFSE) and so are kinetically labile to isomerization and ligand exchange in aqueous solution. By exchanging the iron for chromium(III) which has the same charge, is d^3, and has a significant CFSE, kinetically inert

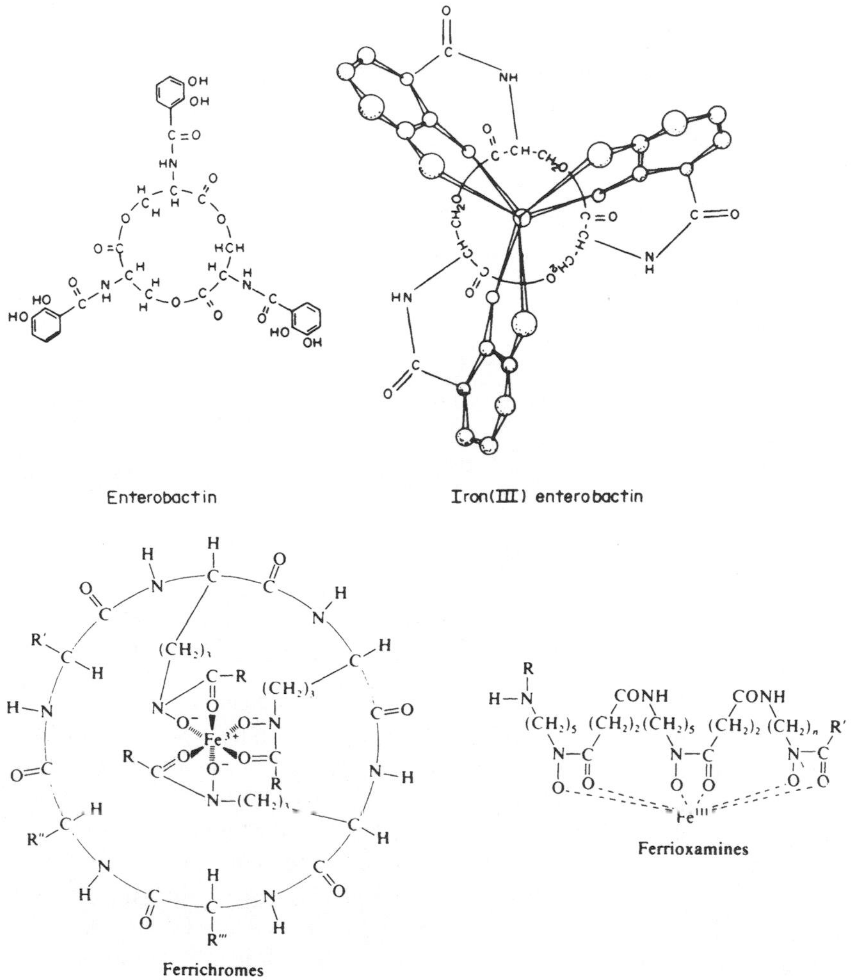

Enterobactin

Iron(III) enterobactin

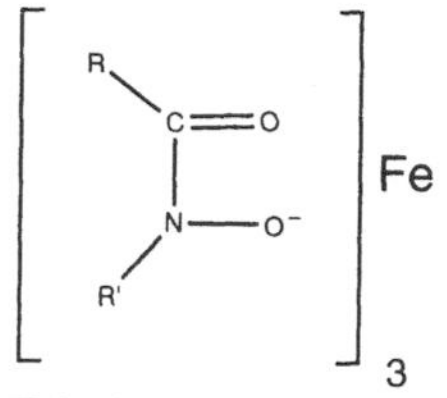

Tris-hydroxamic

2.1

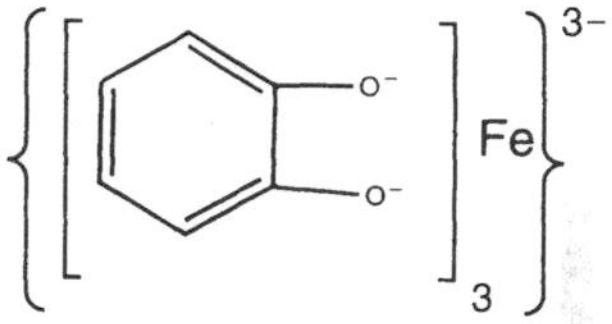

Tris-catecholic

2.2

Ferrichromes

Ferrioxamines

The K_F values cannot readily be compared as they have been measured at different pH values. Transferrin has $K_F = 10^{28} M^{-1}$ at pH 7.4, and enterobactin has $K_F = 10^{52} M^{-1}$ but at pH 11.0. If the latter is corrected to assume measurement at pH 6.0 then it falls to $K_F = 10^{26} M^{-1}$.

Fig. 2.5. Representative siderophores (reproduced with permission from Pergamon Press).

analogues with well characterized d–d transitions can be obtained and studied spectroscopically. By comparing the optical (UV-visible) and circular dichroism (CD) spectra of simple tris-hydroxamate and tris-catecholate chromium(III) complexes with those for the chromium-substituted siderophores the absolute configurations of the siderophores can be established.

The crystal structures of several hydroxamic siderophores including ferrichrome (R = $-CH_3$, R' = R" = R"' = —H), ferrichrysin (R = $-CH=C(CH_3)CH_2CO_2H(trans)$, R' = R" = $-CH_2OH$, R"' = —H and ferrichrome A (R = $-CH_3$, R' = R" = $-CH_2OH$, R"' = —H) have been solved and the total synthesis of a number of siderophores including enterobactin, aerobactin, ferrichrome, and rhodorotulic acid has been achieved. So far it has not been possible to grow crystals of the iron complex of enterobactin. Application of the principle of isomorphous replacement has enabled the retrieval of useful information concerning the metal–ligand interaction.

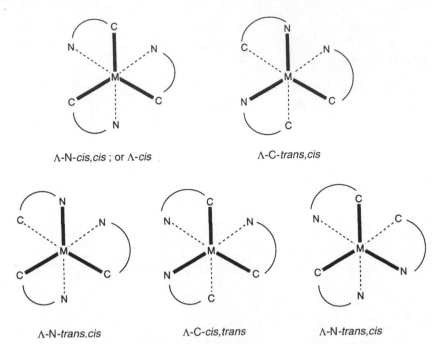

Λ-N-*cis,cis* ; or Λ-*cis*

Λ-C-*trans,cis*

Λ-N-*trans.cis*

Λ-C-*cis,trans*

Λ-N-*trans,cis*

Fig. 2.6. The five Λ-isomers of ferrioxamine B.

$r_{V(IV)} = 0.58\text{Å}$
$r_{Fe(III)} = 0.65\text{Å}$

Vanadium(IV) is slightly smaller than iron(III) and can replace the iron in the enterobactin. The reaction of VO(acetylacetonate)$_2$ with protonated enterobactin (H$_6$–ent) in the presence of KOH leads to the displacement of the vanadyl oxygen atom and the formation of [V(ent)]$^{2-}$. The crystal structure (Fig. 2.7) shows the incorporation of vanadium(IV) and that the enterobactin has a Δ-conformation. The average bond lengths (V–O1A, 1.946Å, and V–O2A, 1.939Å) are essentially the same showing that the binding units are perfectly placed above the serine-derived ring to allow metal coordination. This feature, together with hydrogen bonding between the amidic protons on (N1A) and the catecholic oxygen atoms (O1A), is believed to play a key role in the metal-enterobactin binding and hence in the natural iron(III) complex.

Ligand exchange in the chromium(III) complexes is slow and so they can be used as probes for the mechanism of iron transport. Three different

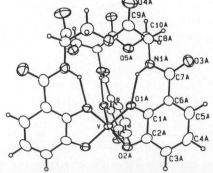

Fig. 2.7. The structure of the vanadium(IV)–enterobactin complex (reproduced with permission from *Angewandte Chemie*).

types of mechanism have been proposed for siderophore activity (Fig. 2.8). In the first, the 'European' route, the siderophore (ferrichrome) transports the metal across the cell membrane to the cell interior where it is released by a non-destructive process and so the ligand is available for reuse. The second is the 'Taxi', where the siderophore (ferrioxamine) delivers the metal to the outer cell membrane surface where it is transferred to a secondary transport device which carries it to the cell interior where it is released. The third mechanism is called, somewhat with tongue in cheek, the 'American' because, the metal is transported across the cell membrane by the sidero-phore (enterobactin) to the cell interior where the complex is broken up, by a hydrolase, so destroying the ligand and giving an example of 'built-in obsolescence'.

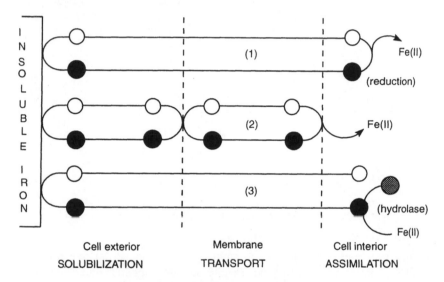

Fig. 2.8 . Siderophore transfer mechanisms.

Many synthetic analogues of particularly the catecholato–siderophores have been prepared (Fig. 2.9), in particular by Kenneth Raymond and his co-workers. This research has been driven both by a need to understand the nature of the catecholate site and by the need to find a chelating device which specifically scavenges iron from the body in cases of iron overload. Iron overload is common in people suffering from the genetic blood disorder β-thalassaemia, also known as Cooley's anaemia, a type of anaemia that prevents the manufacture of haemoglobin, the oxygen-carrying blood protein. It is treated by regular blood transfusions but this has severe side-effects including the accumulation of excess iron causing diabetes, tissue damage, and eventually death.

The siderophore desferrioxamine B (that is the metal-free siderophore) has been used effectively but it causes allergic responses in some patients and can only be used in injections as it is poorly absorbed in the gut. Linear, cyclic, and bicyclic synthetic analogues, representatives of which are shown below, have been synthesized and it is hoped that an alternative to desferrioxamine B will emerge (Section 7.1).

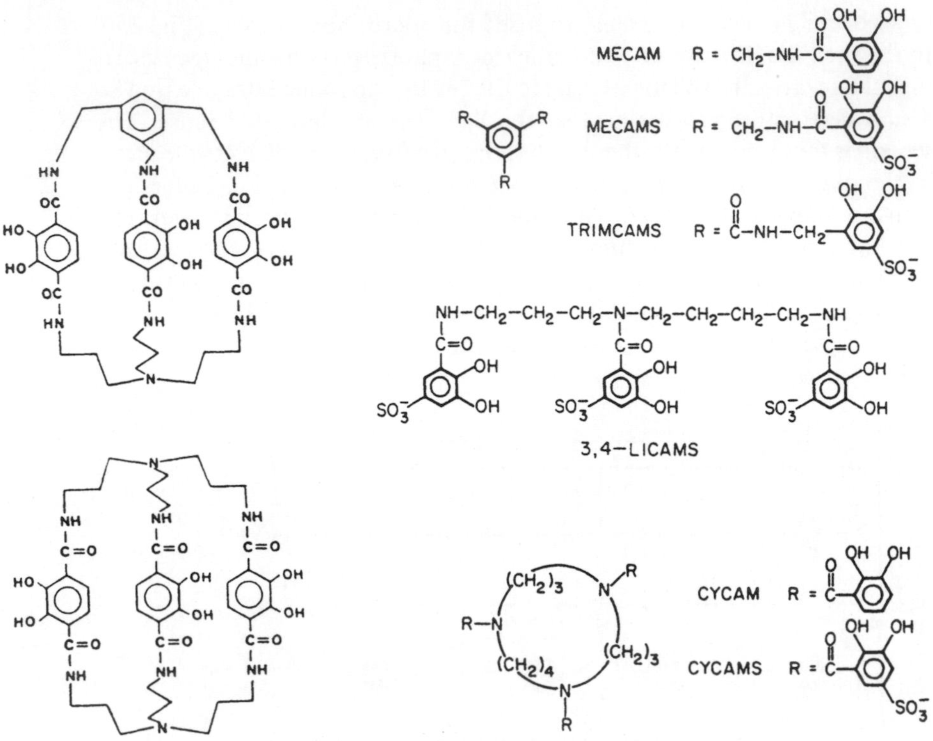

Fig. 2.9. Some synthetic analogues of siderophores.

2.3 Other storage and transport systems

Metallothioneins and phytochelatins

Mammalian metallothioneins are small (~61 amino acids) cysteine-rich (~30%) proteins which can bind seven essential, copper(I) and zinc(II), or non-essential, toxic, cadmium(II) and mercury(II), metals per polypeptide. The biological role is not clear but it is likely that they are involved in metal metabolism—metal storage and transfer, and the detoxification of toxic metals. The metallothionein from the fungus *Neurospora crassa* appears to play a role in the copper metabolism of the fungus.

The use of NMR spectroscopy as a technique for determining protein structure has developed rapidly and has seen application in the study of the structures of metallothioneins. Two-dimensional [^{113}Cd-^{1}H] NMR studies of [Cd$_7$]-metallothioneins in solution, by Kurt Wüthrich and his co-workers, have led to the determination of three-dimensional structures for the metallothioneins from rat liver, rabbit liver, and human liver. All three proteins adopt a similar conformation in solution, and there are two distinct metal clusters present. The seven Cd atoms are bound in clusters of three and four metal atoms held together by terminal and bridging cysteine residues.

The crystal structure of the [Cd$_5$Zn$_2$]-metallothionein from rat liver was solved by David Stout and his research collaborators and confirms the predictions from NMR showing the presence of a [Zn$_2$Cd]-cluster and a [Cd$_4$]-cluster (Fig. 2.10). Each metal atom is tetrahedrally coordinated and the six-membered ring shows distortions from an ideal chair conformation.

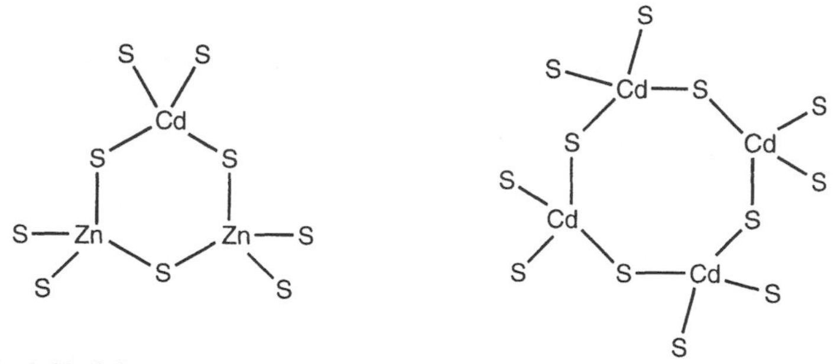

Fig. 2.10. Schematic of the clusters in metallothionein (the cysteine links are represented by S).

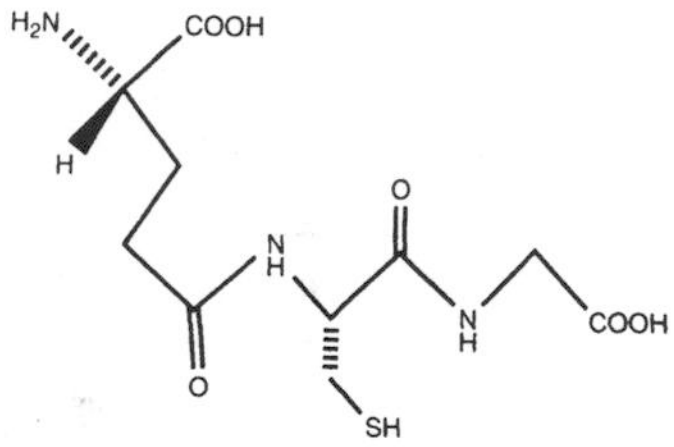

Glutathione: a tripeptide derived from glutamic acid, cysteine, and glycine.

A second class of metal ion detoxificants is a family of sulphur-rich glutathione related peptides bearing multiple γ–Glu-Cys dipeptide units, [(γ–Glu-Cys)$_n$–Glu)] or [(γ–EC)$_n$–G] These occur in plants, and some fungi, and have been termed phytochelatins. A heterogeneous mixture of [(γ–EC)$_n$–G] peptides with n varying from 2–5 is synthesized by the organism in response to the presence of heavy metals such as cadmium(II), copper(I), mercury(II), and lead(II). These peptides bind the metal ion in a metal–cysteine thiolate cluster; acid-labile sulphur is also a component of the Cd-EC complexes. The structure of the metal site is not yet established. Three distinct species have been isolated from fission yeast, *Schizosaccharomyces pombe*, and it has been proposed from spectroscopic studies that two of the species each have six cadmium atoms associated with them together with an inorganic sulphide.

Type I $\{[(\gamma–EC)–G]_x.[(\gamma–EC)_2–G]_x \}[Cd^{2+}]_x$

Type II $\{[(\gamma–EC)_2–G]_6.[(\gamma–EC)_3–G]_2 [S^{2-}]\} [Cd^{2+}]_6$

Type III $\{[(\gamma–EC)_2–G]_2.[(\gamma–EC)_3–G]_3[(\gamma–EC)_4–G]_1[S^{2-}]\} [Cd^{2+}]_6$

Isolates of the Cd–[(γ–EC)] peptide complex from the yeast *Candida glabrata* have been shown to contain [(γ–EC)]-coated CdS crystallites.

Ceruloplasmin

Ceruloplasmin is one of the most enigmatic of copper proteins. It is intensely blue in colour and is associated with the glycoprotein of the α_2-globulin fraction of mammalian blood. Its physiological role is not known with certainty but it is believed to be involved with copper transfer. It has seven, or eight copper atoms present in Type I, Type II, and Type III sites (see 4.1).

Vanadium storage and transport

A class of tunicates, called ascidians (sea squirts), have cellular vanadium contents with concentrations of more than one-million-fold over that of the sea-water in which they live. In ascidians the vanadium is stored in the vacuoles of the blood cells, which are referred to as vanadocytes, as vanadium(III). The vanadium is captured from the sea-water as vanadate and is incorporated in the vacuoles through the phosphate channel taking advantage of the similarity between vanadate (VO_4^{3-}) and phosphate (PO_4^{3-}). Inside the vacuole the vanadium is reduced to the cationic forms V^{3+} and VO^{2+}. It was thought for a long time that the vanadium was responsible for dioxygen carriage, hence the term haemovanadin, and that the pH of the blood was about 1 due to the presence of sulphuric acid. Both of these statements are now believed to be unfounded; the true function of the vanadium is as yet unknown but one possibility is that it could act as a sink for protons and electrons.

Vanadium is extracted from the soil and accumulated by the *Amanita muscaria*, or fly agaric, mushroom as a low molecular weight complex known as amavadin. It can be concentrated by up to 400ppm against background levels of less than 0.1ppm. Ernst Bayer identified the ligand involved as 2,2'-(hydroxyimino)dipropionic acid [HIPDAH$_3$] by use of degradative organic techniques and it was found to form a 2:1 complex with the metal. Bayer was later able to achieve the total synthesis of this compound. The vanadium was shown by EPR to be present as vanadium(IV) and it was initially assumed that this was as the vanadyl centre, V^{IV}O. It was also postulated that amavanadin may be involved in electron transfer processes as it can be reversibly oxidised from vanadium(IV) to vanadium(V).

David Garner and his colleagues have now synthesized and characterized [Δ-V(S,S)-HIDPA)$_2$]$^-$ and established via X-ray and spectroscopic techniques that the oxidized form of amavadin consists of an approximately equal mixture of the Δ- and Λ- forms of this anion. The crystal structure of the anion shows an octacoordinated vanadium(V) atom with each ligand providing two unidentate carboxylato groups and one unusual η^2-N,O bridging interaction (Fig. 2.11). The structure persists in solution and is common to the V^{IV}/V^V reduction–oxidation pair. This feature of the retention of structure throughout a one-electron change is also found at the metal centres in blue copper proteins, cytochromes, and iron–sulphur proteins.

H$_3$C — C(H) — COOH
|
N — OH
|
H$_3$C — C(H) — COOH

2,2'-(hydroxyimino)dipropionic acid
[hidpaH$_3$]

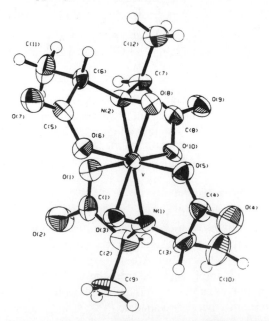

Fig. 2.11. Structure of the anion of [PPh$_4$][Δ-V(S,S)-HIDPA)$_2$]·H$_2$O (reproduced with permission from the American Chemical Society).

3 Dioxygen management— storage and transport

When astronauts venture into space they are reliant upon life-support systems. On earth we exist because we each have our own built-in support system. When we breathe, dioxygen enters our bloodstream through interaction with the iron atoms present in our oxygen-carrier haemoglobin (Hb) and then is carried to the tissues where it is stored by myoglobin (Mb) until it is required for metabolic processes. The iron is bound to a porphyrin ring present in the metalloprotein and this unit is called a haem protein. Hb is a tetrameric metalloprotein having molecular weight of 65,000 and four haem-units; Mb, which is also a haem-protein, may be regarded as a monomer of Hb. The haem unit absorbs one molecule of dioxygen per iron atom and is responsible for the red colour of your blood.

The potency of the haem units may be gauged by comparison of the solubility of dioxygen in water and in blood. In water the solubility is quite low with $6.59 cm^3$ of dioxygen dissolving in $1 dm^3$ at $20°C$ and 1 atmosphere to give a $3 \times 10^{-4}M$ solution. It is possible to dissolve $200 cm^3$ of dioxygen in blood under the same conditions to give a $9 \times 10^{-3}M$ solution. Blood therefore can carry about 30 times as much dioxygen as water.

The dioxygen carriage and storage proteins are called respiratory pigments and are required by all animals and many other species. Not all species, however, use haem proteins. Molluscs and arthropods such as lobsters, squid, cuttlefish, and octopi use haemocyanin, a dinuclear copper protein which absorbs one molecule of dioxygen per two metal atoms and gives the blood a blue colour. Certain marine worms use haemerythrin, a diiron non-haem protein which gives their blood a violet colour.

A porphyrin is a compound having a fully conjugated cyclic structure of four pyrolle rings linked via their 2- and 5- positions by methine (−CH−) links.

A dinuclear protein is one which contains two metal atoms, often in close proximity.

3.1 Haemoglobin and myoglobin

Early X-ray crystal structural studies on Hb by Max Perutz and John Kendrew revealed that the active site in Hb and Mb is the haem which is tightly bound to the protein chain, but the question of whether the dioxygen was bound to the metal in a side-on (**3.1**) or bent, end-on (**3.2**) manner was not solved at this time. Small molecule model compounds for this interaction were devized and played an important role in helping to understand the nature of the Fe–O_2 interaction. Two jigsaw puzzles were completed in parallel with movements in each helping towards the solution of the other.

3.1   **3.2**

Model systems

Extensive studies on the reaction of dioxygen with cobalt(II) complexes of Schiff base ligands [Co(SB)] gave the first real clues to the probable nature

The imine product from the condensation reaction of a primary amine and a carbonyl compound is called a Schiff base (e.g. L').

H_3C ═N   N═ CH_3

H_5C_6 ─O⁻   ⁻O─ C_6H_5

L¹

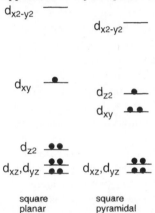

3.3

Compound	Type	O–O(Å)
O_2	O = O	1.21
KO_2	superoxide	1.28
H_2O_2	peroxide	1.49

of the metal–dioxygen interaction. If the reaction of the Co(SB) complex with O_2 was carried out in a coordinating solvent such as dimethyl sulphoxide (DMSO) or in a non-coordinating solvent with the presence of added Lewis base (B) such as pyridine or imidazole, then at temperatures of 0°C or below, rapid and reversible uptake of O_2 corresponding to the formation of 1:1 Co(SB)(B)O_2 (**3.3**) complexes occurred; in the absence of (B) an oxy-bridged dimer was obtained.

The purpose of the strongly coordinating axial ligand may be rationalized by considering the d-orbital energy levels of cobalt(II) in the different co-ordination environments (Fig. 3.1). In the square-planar starting material the unpaired electron is located in the d_{xy} orbital and so is not available for interaction with the approaching dioxygen. The axial ligand leads to a square pyramidal arrangement and raises the d_{z^2} orbital above the d_{xy} level and it is this configuration which appears to be a prerequisite for dioxygen interaction.

$d_{x^2-y^2}$ ——

$d_{x^2-y^2}$ ——

d_{xy} ——

d_{z^2} ——

d_{xy} ——

d_{z^2} ——

d_{xz}, d_{yz} —— d_{xz}, d_{yz} ——

square planar   square pyramidal

Fig. 3.1. Energy levels of the d orbitals showing the d⁷ occupancy of Co(II).

After oxygenation the complexes still have one unpaired electron and the EPR spectra of the oxygenated complexes indicates that this electron has only a small spin density at the cobalt nucleus. This means that the electron is delocalized on the dioxygen ligand and that the cobalt dioxygen interaction should be designated as Co(III)–O_2^-. The X-ray crystal structure of the complex Co(L¹)(DMSO)O_2 showed that the di-oxygen bonding was end-on with a Co–O–O angle of 125°; furthermore the O–O distance of 1.26Å is very close to that of the superoxide anion (O_2^-).

It is useful at this point to remember the molecular orbital energy level diagram for dioxygen (Fig. 3.2) and to add the electrons which generate the O_2, O_2^-, and O_2^{2-} systems. The bond orders resulting are 2, 1.5, and 1 and this is reflected in the increase in bond length. A crystal structure of a dioxygen complex by revealing the O–O distance can therefore help in deducing the nature of the metal–dioxygen interaction.

If one can use these small molecule ligands to try and help understand the interaction of dioxygen with a metal then why not use porphyrins themselves in the modelling of the biosite? The problem that arises is that the simple iron(II) porphyrins react irreversibly with dioxygen in the presence of axial base but this is then followed by irreversible formation of a μ-oxo dimer.

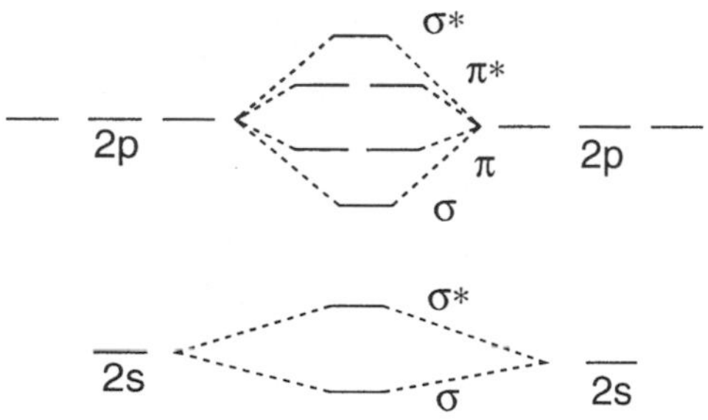

Fig. 3.2. Simplified molecular orbital energy level diagram for dioxygen (omitting the 1s interaction).

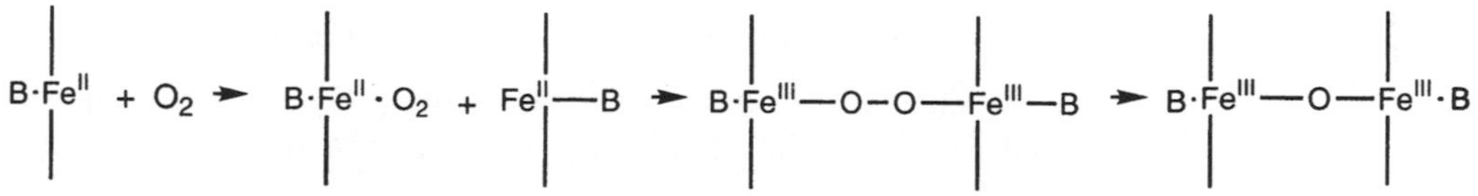

The way in which this problem has been overcome is to work at low temperatures in order to show the reactions leading to dimerization and to introduce steric constraints onto the porphyrin such that dimerization is inhibited. The classic example of the latter is the 'picket fence' porphyrin developed by James Collman and his group, *meso*-tetra(α,α,α,α-*o*-pivalamidophenyl)porphyrin, and many other modifications of porphyrins have subsequently been developed and used to study the Fe-O₂ interaction.

The principle involved is that the porphyrin should be modified in such a way that there is great steric bulk on one side in order to inhibit bimolecular reactions involving two iron centres and dioxygen, and that the iron atom should be five-coordinated by a bulky base on the unhindered side of the porphyrin. This leaves a hydrophobic pocket for interaction with dioxygen (Fig. 3.3).

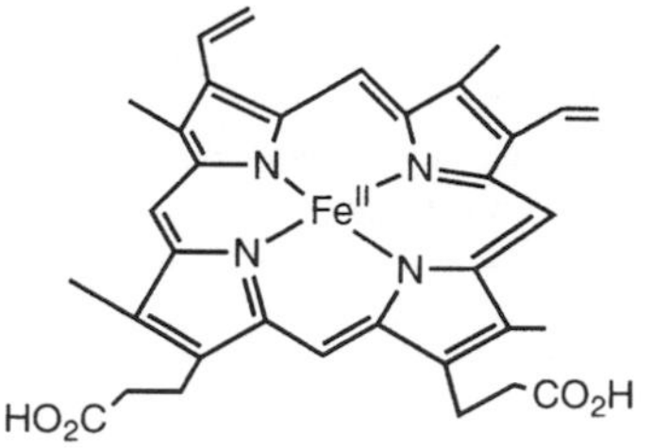

A haem unit
(iron(II)protoporphyrin IX)

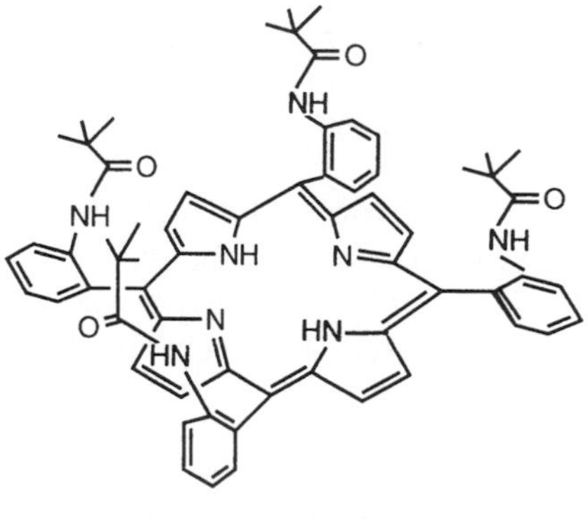

Collman's 'picket fence' porphyrin

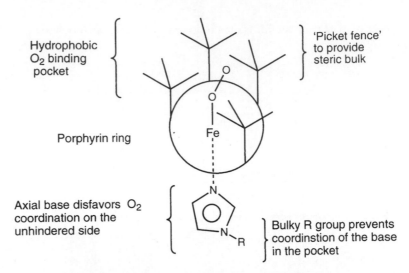

Hydrophobic O_2 binding pocket

'Picket fence' to provide steric bulk

Porphyrin ring

Axial base disfavors O_2 coordination on the unhindered side

Bulky R group prevents coordinstion of the base in the pocket

Fig. 3.3. The binding pocket of the 'picket fence' porphyrin.

This hydrophobic pocket has also been realized by the use of molecular straps and caps, and the axial imidazole has been introduced as a tailbase in conjunction with the pocket.

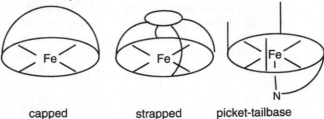

capped   strapped   picket-tailbase

The crystal structure of the iron(II) 'picket fence' porphyrin complex in which 1-methylimidazole is the axial ligand confirmed the bent, end-on geometry of the dioxygen and showed reversible binding. The dioxygen was disordered and so an accurate O–O distance was not available; however, when 2-methylimidazole was used as the axial ligand, a distance of 1.25Å was detected. This and the observation of v_{O-O} 1159cm^{-1} allowed assignment of an Fe^{3+}–O_2^{-} interaction and the resonance Raman spectrum gave an Fe–O_2 stretch at 568cm^{-1}, close to that of 567cm^{-1} in oxyhaemoglobin.

The structures of the natural sites

John C. Kendrew and Max F. Perutz were awarded the Nobel Prize for Chemistry in 1962 for studies on the structures of globular proteins including haemoglobin.

Subsequent to these studies the stereochemistry at the iron sites in both oxy-Mb and oxy-Hb were revealed through their crystal structures, by work carried out by Simon Phillips and Boaz Shaanan respectively. A comparison was now possible with the structures of the deoxy-forms of the proteins. In the deoxy forms the iron atom is five-coordinate, high spin, and sited below the plane of the porphyrin ring towards an axial imidazole ligand. On oxygenation the iron in Hb moves towards the porphyrin plane (from 0.60Å to 0.13Å from the plane) and pulls this histidine with it thus commencing

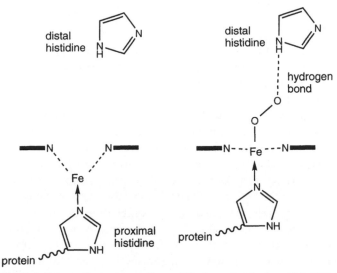

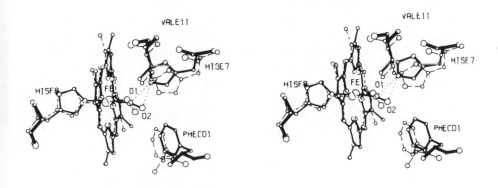

Fig. 3.4. Schematic representations of the iron sites in deoxy-Mb [Hb] (left) and oxy-Mb [Hb] (right).

Fig. 3.5. Stereo view of superimposed haem environments in oxy-Hb (solid lines) and oxy-Mb (dashed lines); the hydrogen bonds to the distal histidine are shown as dotted lines (reproduced with permission from *Nature*).

the cooperativity shown by Hb in binding dioxygen (Fig. 3.4). In Mb this change is not so pronounced (from 0.42Å to 0.18Å). The dioxygen atom is found to be bent, end-on, as in the models, and by using neutron diffraction analysis it was possible to show that the bound dioxygen forms a medium strong hydrogen bond with the distal histidine in oxy-Mb (Fig. 3.5). The Fe–O–O bond angle has been found to vary from 115° in oxy-Mb to ~130° in the 'picket fence' model and 156° in oxy-Hb; these angular changes may reflect the differing steric constraints imposed at the binding site.

The solving of the problem of the nature of the dioxygen–iron interaction in Hb and Mb has been like a detective story with clues coming in from different chemical directions. The usefulness of synthetic model studies as a complementary technique has been demonstrated together with the value of a cooperative approach to problem-solving.

Cooperativity

The most significant property of Hb is cooperative oxygen binding in which the oxygen affinity of the tetramer rises with increasing oxygen saturation. The four subunits of Hb do not act independently and appear to communicate with each other such that when one haem unit interacts with an oxygen molecule the other three units show an enhanced ability to interact with dioxygen. Hb is less efficient than Mb at oxygen uptake under low oxygen pressures and so in muscle tissue where this is so, there is a thermodynamically favourable transfer of dioxygen from oxy-Hb to oxy-Mb. The basis of cooperativity has been proposed to arise from a 'trigger' mechanism. In deoxy-Hb the high-spin Fe(II) atom is about 0.60Å below the mean plane of the porphyrin ring and upon oxygenation the iron atom changes to low spin and moves into the plane of the ring. At the same time the proximal histidine is moved by about 0.8Å so causing conformational changes in the protein which can in turn induce changes at the remaining iron sites in a concerted action.

'Synthetic Blood'

One interesting potential application of synthetic porphyrin analogues of Hb is in the generation of synthetic blood. Eishun Tsuchida and his co-workers found that by incorporating phospholipid chains onto the framework of the meso-5,10,15,20-tetra-(o-$\alpha,\alpha,\alpha,\alpha$-pivalamidophenyl) porphyrinatoiron(II) complex (Fig. 3.6) it was possible to investigate the oxygen-carrying potential of the lipophilic system when embedded in a phospholipid bilayer

Lipids are triglycerides of fatty acids; a phospholipid is derived from glycerol by esterifying two of the hydroxyl groups with fatty acids and the third by a derivative of orthophosphoric acid. A lipophilic substance has a strong affinity for lipids.

$$CH_2OC(=O)R$$
$$R'C(=O)OC$$
$$C-O-\overset{\overset{O}{\|}}{P}-OR''$$
$$O^-$$

Cell membranes are about 70Å thick and are composed of proteins and phospholipids. The phospholipids form a bilayer (Fig. 3.6) in which the long hydrocarbon chains point inwards leaving the polar phosphate entities on the surface. The proteins both coat the surface of this bilayer and bridge across it. This has been termed the 'Fluid Mosaic Model' of the membrane.

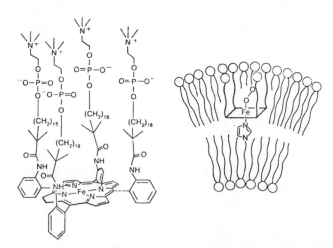

Fig. 3.6. An iron–porphyrin complex reversibly binds dioxygen when embedded in a phospholipid bilayer (reproduced with permission from the American Chemical Society).

related to that found in a cell membrane. The best of the complexes prepared binds and releases dioxygen in response to changes in dioxygen partial pressure in a manner similar to red blood cells, and can take up about 20 times as much oxygen as an equivalent amount of water which is again similar to blood. However, for such a substance to be used as a substitute for haemoglobin it must be physiologically acceptable and that, as yet, has not been proved.

3.2 Haemerythrin and haemocyanin

Although the dioxygen storage proteins in marine invertebrates such as sipunculids (peanut worms) and annelids (segregated worms) and the carriers in several species of the phyla Mollusca and Arthropoda are called haemerythrins (Hr) and haemocyanins (Hc), respectively, they are not actually haem-proteins. Haemerythrin is a dinuclear iron compound, and haemocyanin is a dicopper compound, neither of which have porphyrin rings present.

Oxy- and deoxy-haemerythrin

X-ray crystal structures of the active site in haemerythrins have been carried out by Ronald Stenkamp's research team and reveal an unusual structure in which the two iron atoms are triply bridged by an oxo-atom, the carboxylate functions from a glutamate residue and an aspartate residue, such that the bridges provide the central ligands of a confacial bioctahedral array. Five of the remaining ligands are provided by histidine ligands from the polypeptide chain, leaving one accessible ligand site. The presence of the Fe–O–Fe motif in haemerythrins is consistent with the strong antiferromagnetic coupling ($-J = 134 \text{cm}^{-1}$) and the Fe–O–Fe vibration mode at 510cm^{-1} in the six-coordinate *met*-forms which have iron(III) present and are inactive towards dioxygen. The Fe–O–Fe motif has now been found to be widespread in nature, also occurring in a range of metalloenzymes such as ribonucleotide reductase, methane mono-oxygenase, and the purple acid phosphatases.

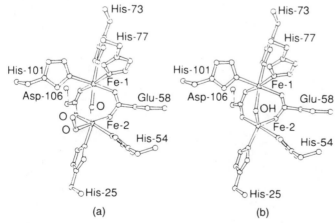

Fig. 3.7. The structures of (a) oxy- and (b) deoxy-haemerythrin (reproduced with permission from the Royal Society of Chemistry).

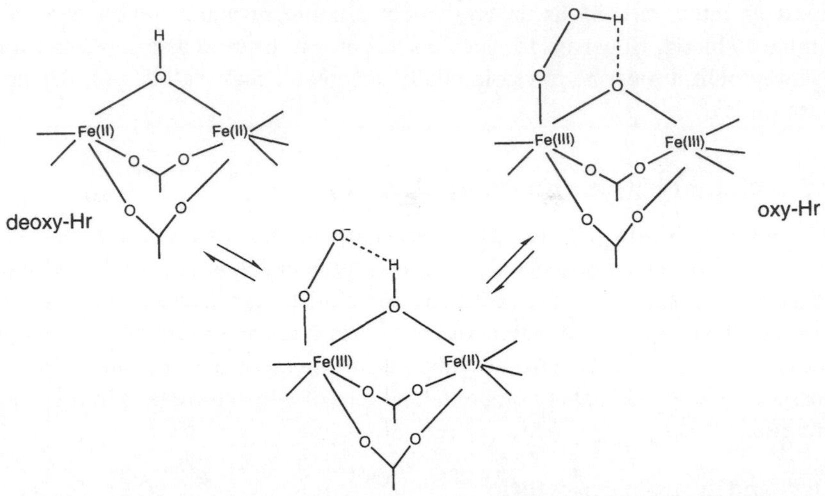

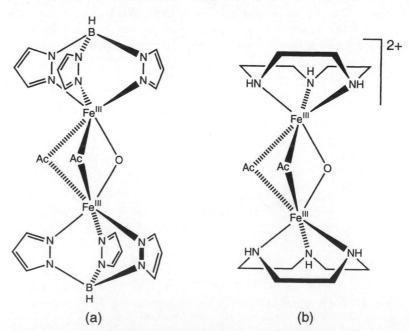

Fig.3.8. A mechanism for the uptake and release of dioxygen from haemerythrins.

In deoxy-Hr the iron is present as iron(II) and the oxo-bridge is from a hydroxyl group (Fig. 3.7). One of the iron(II) atoms is six-coordinate and the second is five-coordinate and so coordinatively unsaturated; the intermetallic separation is 3.3Å . This site is occupied in oxy-Hr by a dioxygen bound as a hydroperoxide group hydrogen bonded to the central oxo-atom. A mechanism for the uptake and release of dioxygen is shown in Fig. 3.8.

The Fe–O–Fe unit acts as an acid–base centre and the dioxygen binds at the vacant site at one iron(II). One-electron transfer then occurs to the

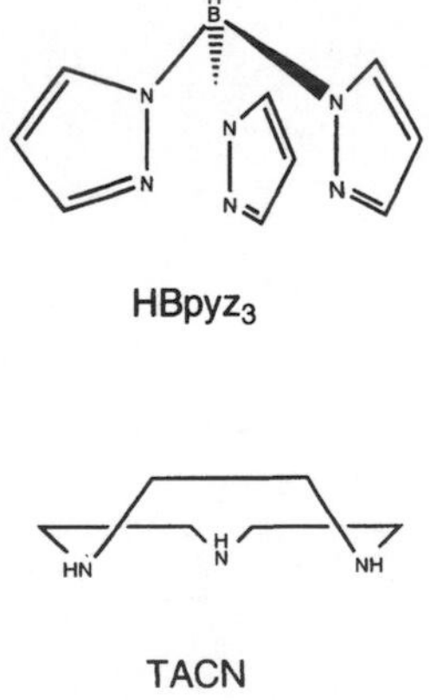

HBpyz₃

TACN

(a)

(b)

Fig. 3.9. Model complexes for the haemerythrin Fe–O–Fe core derived by (a) Lippard and (b) Wieghardt.

dioxygen followed by hydrogen abstraction from the bridging hydroxyl to the bound superoxide. This is followed by electronic rearrangement within the dinuclear unit. The facile proton transfer allows electrons to flow in either direction at the active site.

Modelling studies for hemerythrins are necessarily corroborative and have mainly concerned the six-coordinate *met*-forms. The ligands involved have been derived from tridentate nitrogen ligands such as the tripodal hydridotris(1-pyrazoyl)borate anion (HBpyz$_3$) and the macrocycle 1,4,7-triazacyclononane (TACN) which provide a rigid face-capping donor set and so facilitate the formation of confacial bioctahedral complexes (Fig. 3.9). The model complexes were developed in parallel by the groups of Stephen Lippard and Karl Wieghardt and the complexes prepared using a technique, now termed 'spontaneous self-assembly', in which all of the components are simply mixed and left to crystallize. These ligands have proved to be versatile and have also been used to cap one face of a tetrahedron in the derivation of models for the active site in zinc enzymes.

Oxy- and deoxy-haemocyanin

The speculative modelling of the active site in oxy-haemocyanin provides another detective story which is well worth private study. From cumulative spectroscopic studies on the metalloprotein a site proposal (**3.4**) was made in which the two copper atoms were coordinated by three histidine ligands each together with an endogenously bridging donor atom, believed to be an oxygen from tyrosinase, and an exogenously bridged *cis*-1,2-peroxide.

3.4

This proposal generated many model studies and the thrust of these studies was to provide an endogenous bridge, as this was held responsible for the observed strong antiferromagnetic coupling between the two copper(II) ions and to replicate this together with the intercopper distance of *c*.3.5Å, and the charge transfer band at 330nm. Much of the seminal work on model complexes capable of activating dioxygen derives from the laboratory of Kenneth Karlin. The eventual solution of the crystal structure of deoxy-haemocyanin from the spiny lobster (*Panulirus interruptus*), by Wim Hol and his group (Fig. 3.10a), confirmed that the dinuclear nature of the site and the presence of three histidine ligands per copper atom but eliminated the concept of an endogenous bridge as there was no conserved candidate suitable for such bridging within 12Å of the site.

One intriguing model (**3.5**) in which there was no endogenous bridge eventually emerged from the laboratory of Nobumasa Kitajima. The terminal ligands were provided by the 3,5-diisopropyl derivative of the tripodal ligand

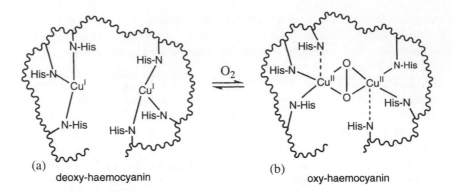

Fig. 3.10. Schematic representations of (a) the dicopper(I) site in deoxy-haemocyanin and (b) the dicopper(II) site of oxy-haemocyanin.

HBpyz3 and the structure of the complex showed that the dioxygen present had inserted between the copper atoms in a novel μ-η^2:η^2 bridging mode. The spectroscopic and magnetic properties of this molecule closely resembled that of the biosite, so offering a new candidate for the binding of dioxygen at the dinuclear site.

3.5

The dilemma concerning the actual mode of bonding was resolved with the elucidation of the crystal structure, by Karen Magnus, of oxy-haemocyanin from the horseshoe crab (*Limulus polyphemus*). The μ-η^2:η^2 bridge was shown to be present and each copper atom had three histidines associated with it (Fig. 3.10). It is interesting to note that in the three oxy-protein structures so far solved, dioxygen has been found bound to the metal centre in three different ways.

4 Electron transfer

Electron-transfer reactions are central to many of the metabolic processes necessary for the survival of all organisms. Such reactions depend upon the approach of an electron donor and electron acceptor and are classified into two types:

- *inner-sphere electron transfer* in which the coordination spheres of the reactants share a ligand transitorily and so form a bridged intermediate;
- *outer-sphere electron transfer* in which the coordination spheres of the reactants remain intact.

In biology it is the latter process which is most often used and biological electron transfers have been divided into two categories:

- *intramolecular electron transfer,* which occurs at fixed sites within a single protein;
- *intermolecular electron transfer,* which occurs between sites on different proteins.

The latter process leads to electron-transfer chains which function as a series of consecutive electron-transfer reactions between pairs of proteins. Typically electrons are transferred between metal sites that are arranged within a protein, or complex of proteins, such that the electron might need to traverse distances of up to 30 Å via the intervening peptide units.

The reactions of electron-transfer proteins are usually examined in terms of Marcus theory which correlates the rate constant of the reaction (k_{12}) with electron self-exchange rate constants of the reactants (k_{11} and k_{22}) and the equilibrium constant for the electron-transfer reaction k using the equation

$$(k_{21})^2 = k_{11}.k_{22}.K.f \qquad f \approx 1$$

The product $k_{11}.k_{22}$ reflects the intrinsic barrier to electron transfer and K is a measure of the overall reaction free energy ΔG^o. The structure of the protein—the intersite separation, the nature of the intervening groups, and the orientation of the donor and acceptor sites—can play a rôle in determining the reaction rate constant and different types of proteins can have rate constants that vary by several orders of magnitude.

4.1 Blue copper proteins

The copper(II) atoms present at copper-containing biosites have been classified according to their spectroscopic properties: Type I, or blue, has an intense absorption in the visible region ($\varepsilon > 3000M^{-1}cm^{-1}$ at 600nm) arising from S(cysteine) $\rightarrow$ Cu(II) charge-transfer and an electron paramagnetic resonance (EPR) spectrum with an unusually narrow hyperfine splitting ($A_{||} < 0.95 \times 10^{-4}cm^{-1}$) due to an asymmetric environment at the metal; Type II, or normal, has limited absorption and an EPR spectrum typical of small

Henry Taube was awarded the Nobel Prize for Chemistry in 1983 for his research into the mechanisms of electron-transfer reactions of metal complexes and Rudolph Marcus was awarded the Nobel Prize for Chemistry in 1992 for his theoretical work on electron transfer between molecules.

Some standard redox potentials for copper proteins at pH 7.0.

Protein	Redox potential (mV)
Laccase	
Type I	+7854
Type II	+500
Type III	+400(?)
Azurin	
Type I	+330
Plastocyanin	
Type I	+370
Tyrosinase	
Type III	+370

The potentials are much higher than those of aqueous copper ions (+170mV) indicating that Cu(I) is bound more strongly than Cu(II).

Hyperfine splittings result from couplings between the spin of the unpaired electron and that of a neighbouring nucleus. An interaction of an unpaired electron with a nucleus of spin I results in 2I+1 lines of equal intensity and spacing (e.g., Cu(II) [I = 3/2] gives 4 lines). The parameter A is called the hyperfine splitting constant. It reflects the electron structure of the complex and is determined experimentally from the spacings of the lines.

molecule copper(II) complexes($A_{||} > 140 \times 10^{-4} \text{cm}^{-1}$); Type III, which has a strong absorption in the near UV region ($\lambda_{max} = 330 \text{nm}$) and no EPR signal, is believed to consist of a pair of antiferromagnetically coupled copper(II) ions.

The blue copper proteins of which plastocyanin, found in higher plants and green algae, and azurin, found in denitrifying bacteria, serve as representative examples are involved in electron-transfer processes. Plastocyanin is involved in electron transfer between Photosystems I and II in the photosynthetic chain, and azurin in respiratory chains where the role is to transfer electrons between cytochrome c_{551} and cytochrome oxidase.

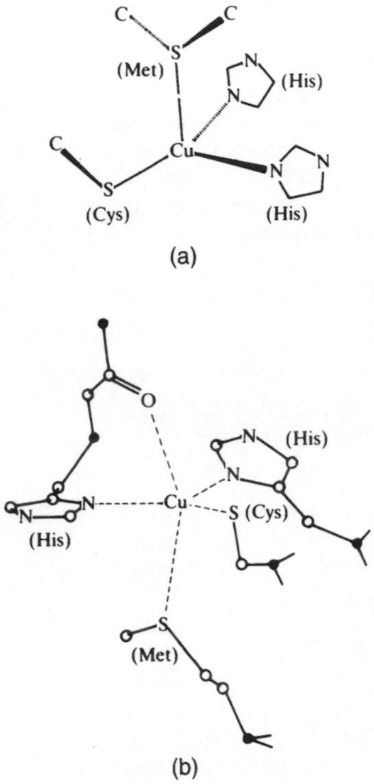

Electron transfer proceeds via transition state structures which are intermediate between those of the reactant and product and, as indicated by the Franck–Condon principle, does not occur until the metal centre is vibrationally excited to a geometry appropriate for that of the product complex in the redox reaction. This can generally be accomplished by a simple adjustment in the bond lengths but for copper it can be more complicated as copper(I) and copper(II) have different preferred geometries—tetrahedral and tetragonal respectively. Any interchange between these would be energetically demanding and so the availability of an intermediate geometry leading to a low reorganization energy is desirable.

Solution of the crystal structures of plastocyanin (Hans Freeman) and azurin in both oxidation states (Edward Baker) (Fig 4.1) has revealed that minimal structural change occurs on reduction of copper(II) to copper(I) (Table 4.1) thus supporting the view that the protein structure is providing a copper site optimized for fast electron transfer. For example, in the azurin structures there is a slight expansion of the copper site on reduction which parallels the expected increase in copper atom radius; the principal copper ligands (S and N) are a compromise between those favoured by copper(I) and copper(II) and the trigonal coordination geometry, with weakly bound axial groups, is intermediate between geometries that might be favoured by copper(I) (trigonal planar) and copper(II) (trigonal bipyramidal). In plastocyanin a distorted tetrahedral geometry with an unusually long copper–methionine interaction is found for each oxidation state.

Fig. 4.1. The copper(II) sites in (a) plastocyanin and (b) azurin (reproduced with permission from the American Chemical Society).

Table 4.1. Bond lengths for the copper sites in plastocyanin and azurin

Plastocyanin	Cu(I)	Cu(II)	Azurin	Cu(I)	Cu(II)
Cu–N–His[37]	2.12	2.04	Cu–O–Gly[45]	3.22	3.13
Cu–S–Cys[84]	2.11	2.13	Cu–N–His[46]	2.13	2.08
Cu–N–His[87]	2.25	2.10	Cu–S–Cys[112]	2.26	2.15
Cu–S–Met[92]	2.90	2.90	Cu–N–His[117]	2.05	2.00
			Cu–S–Met[121]	3.23	3.11

Blue copper sites also exist in multicopper oxidases such as laccase, ceruloplasmin and ascorbate oxidase in which all three site types are present. These enzymes catalyse the four-electron reduction of dioxygen to water with an accompanying one-electron oxidation of the reducing substrate. Cumulative spectroscopic and azide binding studies on laccase from the oriental lacquer tree, *Rhus vernicifera,* led Edward Solomon to propose that the Type II and Type III centres defined a trinuclear copper cluster (Fig. 4.2). The long absence of crystallographic information concerning this type of site was redressed with the solution of the structure of oxidized ascorbate oxidase, from green zucchini squash, by Albrecht Messerschmidt and co-workers, which confirmed the presence of all three types of copper binding site—one Type I, one Type II, and a Type III pair.

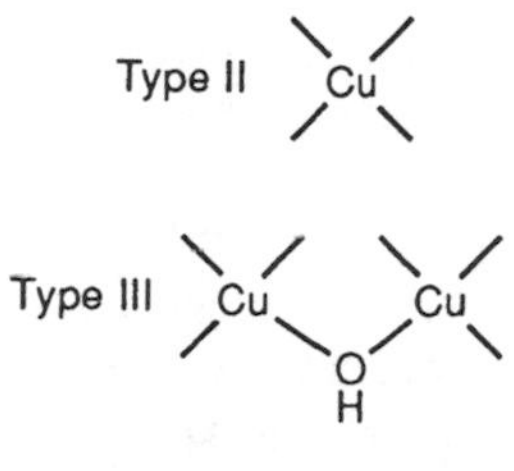

Type II

Type III

Fig. 4.2. Site prediction for the trinuclear cluster in laccase.

The Type I site is closely analogous to that determined for plastocyanin (Cu–S–Cys[508], 1.9; Cu–N–His[446], 2.2; Cu–N–His[513], 2.2; Cu–S–Met[518], 2.9Å) and is 12.2Å distant from the Type III pair. This pair together with the Type II copper atom form a triangular array of copper atoms (Fig. 4.3). The role of the Type I copper is to receive electrons from ascorbate and to transfer them to the triangular array which is the site of dioxygen interaction. Dioxygen binding within the triangular site would allow close interaction and rapid electron transfer to occur from all three copper atoms. Any negative charge which developed at the dioxygen, or oxygen intermediates, would be balanced by the metal ions and the protein so protected from oxygen radicals.

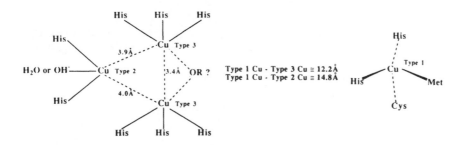

Fig. 4.3. The copper sites in ascorbate oxidase (reproduced with permission from Elsevier).

4.2 Iron–sulphur proteins

Iron–sulphur (Fe–S) proteins are non-haem proteins which are found in all living organisms where they are involved in a wide range of electron-transfer processes. They are also involved in important oxidation–reduction processes such as nitrogen fixation and electron transfer in mitochondria. The iron in these proteins is uniquely bound by sulphur atoms, either from cysteinyl residues present in the protein or by inorganic sulphide. Crystallographic studies have shown that the Fe–S proteins fall into a number of structural types depending upon the numbers of iron and inorganic sulphur atoms present and can be generally designated as [nFe–mS*] where S* represents sulphide.

Rubredoxins [1Fe–0S*]

Rubredoxins, found in bacteria, are the simplest type of Fe–S protein and have one iron atom bound to four cysteinyl residues in a distorted tetrahedral manner (Figs 4.4a, 4.5). The Fe–S bond lengths in the iron(III) form are between 2.24 and 2.33Å and are increased slightly by 0.05Å in the reduced iron(II) form; rubredoxin can be considered as being in an entatic state. The rubredoxins can act as one-electron donor–acceptors.

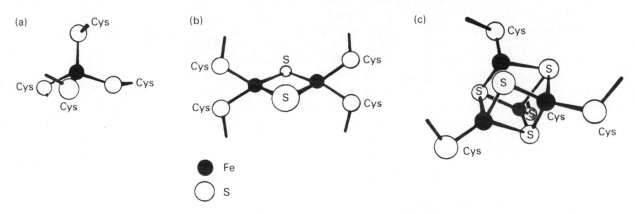

Fig. 4.4. The coordination in Fe–S proteins: (a) rubredoxin; (b) 2Fe–2S* ferredoxins; (c) 4Fe–4S* ferredoxins. S represents sulphide (S^{2-}) and Cys represents cysteinate sulphur atoms (reproduced with permission from Oxford University Press).

[2Fe–2S*] proteins

[2Fe–2S*] proteins have been isolated from mammalian, plant, and bacterial sources and possess a dinuclear structure which consists of two tetrahedra sharing a common edge provided by bridging inorganic sulphur atoms (Figs 4.4b, 4.5). The remaining coordination sites are occupied by cysteinyl residues. The iron–iron separation in the oxidized form is 2.69Å. The cluster transfers and accepts only one electron and on one-electron reduction the added electron is localized on one iron atom to give a mixed valence [Fe(II)–Fe(III)] cluster. In the diiron(III) form the two high-spin d^5 ions are strongly antiferromagnetically coupled giving a diamagnetic cluster;

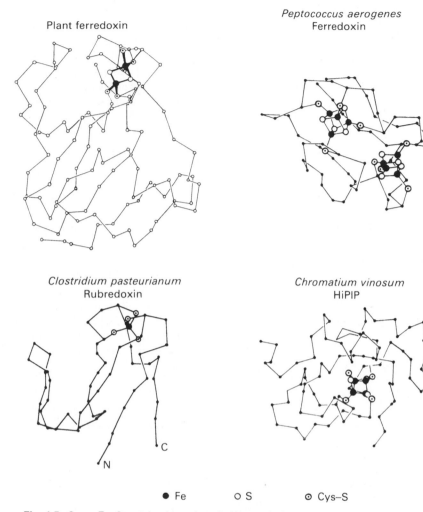

Plant ferredoxin

Peptococcus aerogenes Ferredoxin

Clostridium pasteurianum Rubredoxin

Chromatium vinosum HiPIP

● Fe   ○ S   ⊙ Cys–S

Fig. 4.5. Some Fe–S proteins (reproduced with permission from Oxford University Press).

Mitochondria are the 'powerhouses of cells' enzymatically converting fuels such as glucose into carbon dioxide, water, and energy; the mitochondria are bounded by membranes and contain a large number of internal membranes.

the mixed valence form gives EPR parameters consistent with a coupled high-spin d^5–high-spin d^6 pair thus providing a useful means of differentiation.

A second distinct subclass of [2Fe–2S*] proteins with unique structural features has been isolated from bovine mitochondria and the photosynthetic $b_6 f$ complex and termed Rieske proteins after their discoverer. The higher reduction potentials associated with these iron centres and differences in the Mössbauer and EPR spectra suggested that the overall ligation was different and EXAFS studies have suggested that there is a nonsymmetrical structure with one iron linked to two cysteinyl residues and the second iron linked to two histidine residues (**4.1**).

Cys–S, S, N–His / Fe Fe / Cys–S, S, N–His

4.1

[4Fe–4S*] proteins

The [4Fe–4S*] proteins are cubic with the irons occupying alternate corners of the cube with triply bridging sulphides occupying the other corners. The irons are also ligated by cysteinyl residues (Figs 4.4c, 4.5). The ferredoxin from *Peptococcus aerogenes* contains two [4Fe–4S*] clusters separated from each other by 12Å and has a redox potential of –400mV whereas the high potential protein (HiPiP) from the bacterium *Chromatium vinosum* has a single cluster with a redox potential of +350mV (both potentials are against the standard calomel electrode). In biological systems, there are three available cluster oxidation levels but in any given system only one pair is employed.

The Fe–S clusters may be regarded as mixed valence species. The 2Fe–2S* clusters have localized valencies but the 4Fe–4S* clusters have valence delocalized structures in all cluster oxidation levels, e.g., the iron in a [4Fe–4S*]$^{2-}$ cluster has an average oxidation state of 2.5. In the adjacent scheme localized valencies are used to illustrate the one-electron changes that occur.

$$[Fe_4S_4(SR)_4]^- \rightleftharpoons [Fe_4S_4(SR)_4]^{2-} \rightleftharpoons [Fe_4S_4(SR)_4]^{3-}$$

3Fe(III)1Fe(II) 2Fe(III)2Fe(II) 1Fe(III)3Fe(II)

HiPiP-ox HiPiP-red

Fd-ox Fd-red

No single ferredoxin is yet known to undergo reversible electron transfers between all three oxidation levels and as the structures of the cubes in the clusters are virtually identical it is believed that it is the protein which controls the redox pattern but whether this is a specific structural phenomenon or a general effect is not known.

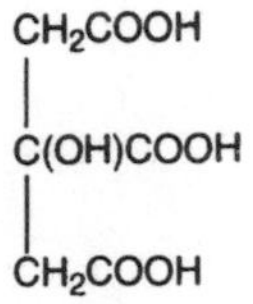

[3Fe–4S*] proteins

This class of Fe–S proteins has been confirmed after controversy over the structure which is now crystallographically characterized as being similar to that of the Fe$_4$S$_4$ unit but with one iron removed to leave a voided cuboidal cluster (**4.2**).

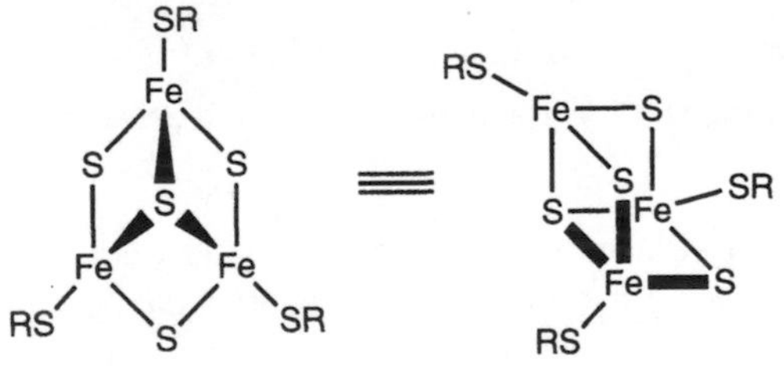

CH$_2$COOH
|
C(OH)COOH
|
CH$_2$COOH

citric acid

[2-hydroxy-1,2,3-propane tricarboxylic acid]

CH(OH)COOH
|
CHCOOH
|
CH$_2$COOH

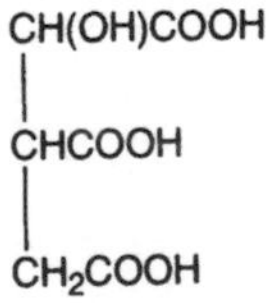

isocitric acid

[1-hydroxy-1,2,3-propane tricarboxylic acid]

4.2

First found in the bacteria *Desulfovibrio gigas* and *Azobacter vinlandii* this cluster is also present in the inactive form of pig heart aconitase, an enzyme which converts citrate to isocitrate by a stereospecific dehydration–rehydration reaction. 3Fe–4S* clusters possess the ability to undergo interconversion reactions to 4Fe–4S* clusters with minimal structural change particularly if there is a labile non-thiol fourth ligand available in the 4Fe–4S* form as in aconitase which has a water molecule present.

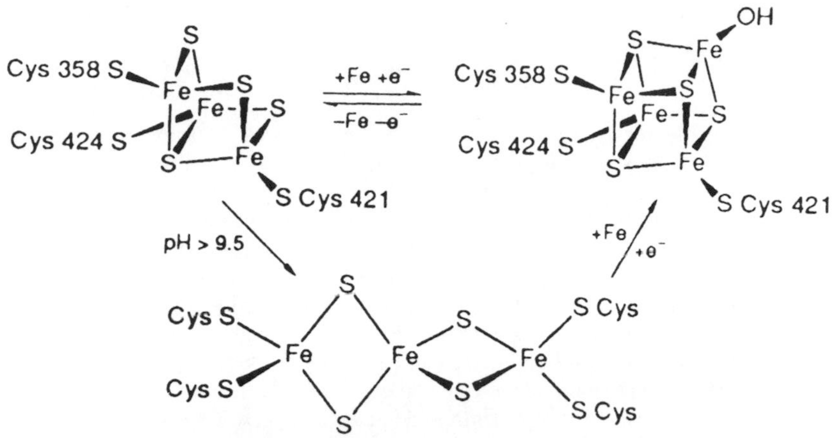

Fig. 4.6. Cluster interconversion in aconitase (reproduced with permission from the *European Journal of Biochemistry*).

The cuboidal 3Fe–4S* form of aconitase is inactive but can be reactivated by the addition of Fe(II) under anaerobic conditions followed by interconversion to the active cuboidal 3Fe–4S* centre (Fig. 4.6). The role of the trinuclear clusters is open to speculation.

Synthetic models for Fe–S proteins

The modelling of Fe–S clusters, in particular by Richard Holm and his research group, opened up a new area of inorganic chemistry. Until the early 1970s the only such cluster that had been prepared, $[\eta^5\text{-Cp}]_4\text{Fe}_4\text{S}_4$ (η^5-Cp is the pentahaptocyclopentadienyl anion), came from organometallic chemistry and was regarded as a laboratory curiosity. During the 1970s an elegant series of experiments established the synthetic technology required to prepare corroborative models for the Fe–S clusters (Fig. 4.7). In the case of the

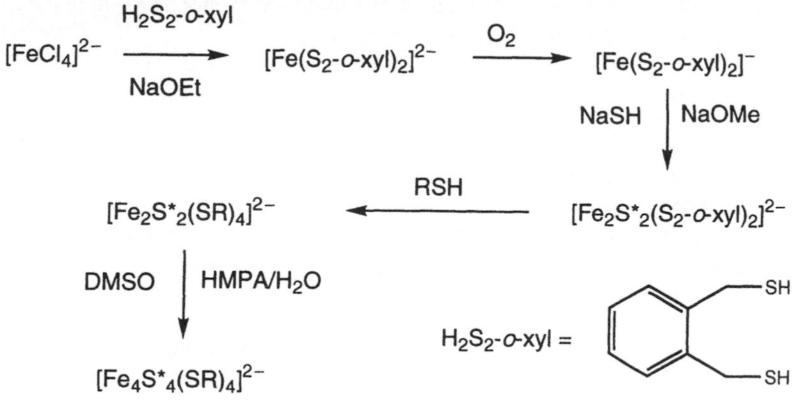

Fig. 4.7. Synthetic routes to model Fe–S clusters (HMPA = hexamethylphosphoramide).

$$[Fe_2S_2(SR)_4]^{4-} \rightleftharpoons [Fe_2S_2(SR)_4]^{3-} \rightleftharpoons [Fe_2S_2(SR)_4]^{2-}$$

1Fe(II)1Fe(II)   1Fe(III)1Fe(II)   1Fe(III)1Fe(III)

Fd-red   Fd-ox

2Fe–2S* clusters the structure for the biosite was anticipated from the structure of the synthetic analogue. In these models thiolate ligands mimic the cysteinyl residues and a range of clusters have now been reported in which glycylcysteinyl oligopeptides are the terminal ligands. Water-soluble clusters such as $[Fe_4S*_4(SCH_2CH_2CO_2)_2]^{6-}$ are also known. At this point, however, there is no model for the 3Fe–3S* cluster. Electrochemical studies on the model systems have established the electron-transfer processes available and reinforce the studies on the natural systems. For example, in the 2Fe–2S* case the above sequence can be detected but nature chooses only to operate at the right-hand end of the series. The facile thiol exchange reactions at the labile tetrahedral iron centres, depicted in Fig. 4.7 with the conversion of $[Fe_2S*_2(S_2\text{-}o\text{-xyl})_2]^{2-}$ to $[Fe_2S*_2(SR)_4]^{2-}$, offers the possibility of both the reconstitution of apoproteins and of the extrusion of intact Fe–S clusters from metalloproteins. In the latter case judicious choice of thiol can be made to introduce spectroscopic probes, for example, $p\text{-CF}_3\text{-C}_6\text{H}_4\text{-SH}$ as an ^{19}F NMR probe. The technique of core extrusion has proved to be a valuable tool in the study of Fe–S and related clusters.

$$\text{Holoprotein} + \text{RSH} \xrightarrow[\text{'unravelling' solution}]{\text{DMSO/H}_2\text{O}} \text{Apoprotein} + \text{cluster}$$

4.3 Cytochromes

Cytochromes are haem proteins which act as one-electron carriers by shuttling between iron(II) and iron(III) at their active site. They are found in all forms of aerobic life and are present, for example, in plant chloroplasts where they participate in photosynthesis and in mitochondria where they take part in the reverse process of respiration.

More than fifty cytochromes have been characterized. Their classification is complex because they differ from one organism to the next, and even from one type of cell to another in the same species. A cytochrome can be tailored to meet the needs of the electron-transfer scheme in which it is involved by variation in the ligands present and of the operating redox potential. Nevertheless four different groups of cytochromes have been defined:
- cytochromes a, in which the haem has a formyl group present;
- cytochromes b, in which the protein is not covalently bonded to the haem;
- cytochromes c, which have covalent links between the haem and the protein;
- cytochromes d, which have the dihydroporphyrin group present.

Cytochromes b and c generally have two strong field axial ligands and so are low spin. They are six-coordinate and so, unlike the haems used in oxygen transport and enzyme activity, have no position for further coordination and so only interact by an electron-transfer mechanism.

In 1930 Keilin realized that the cytochromes consisted of three spectroscopically different components which he designated *a*, *b*, and *c* depending on the position of their absorption maxima. Cytochrome *a* absorbs at 580–590nm (lowest energy); then cytochrome *b* at 550–560nm and cytochrome *c* at 548–552nm (highest energy).

In plants photosynthesis takes place within highly specialized organelles called plastids. In green plants the plastids contain chlorophyll and are called chloroplasts.

Cytochrome *c*, isolated from mitochondrial membranes where its role is to transfer electrons from cytochrome c_1 to cytochrome *c* oxidase, is perhaps the most widely studied of the cytochromes (Fig. 4.8). The mitochondrion is the cellular apparatus that transfers carbohydrate-derived electrons to dioxygen and uses the chemical potential of the reaction to drive the

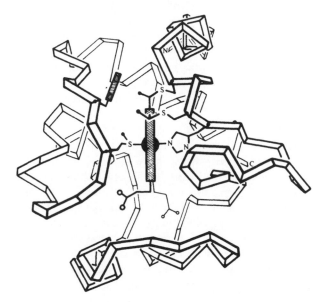

Fig. 4.8. Schematic drawing of the structure of oxidized horse heart cytochrome *c* (reproduced with permission from Harper Collins).

phosphorylation of ADP to ATP. The electron flow occurs through a series, or chain, of several cytochromes (Fig. 4.9). The crystal structures of the iron(II) and iron(III) forms of tuna cytochrome *c* have been solved and are very similar suggesting that there is no conformational change during the redox process. The haem iron is further coordinated by a histidine residue and to the sulphur of a methionine. Electron transfer from the reductase and to the oxidase occurs over a fairly large distance; it may occur through the edge of the porphyrin as if there is a complex formed then the edge-to-edge distance would be about 16Å and relatively rapid electron transfer can occur between metal centres separated by distances of over 10Å.

Some cytochrome *a* types are five-coordinate in contrast to cytochrome *c*. They account for the unusual toxicity of the cyanide ion which can bind strongly to the sixth position at the iron(III) and stabilize it such that it is not reduced. This stops the electron shuttling along the chain. Cyanide can also bind to haemoglobin at the dioxygen site; inhibition of cytochrome *a* activity is, however, more serious than the inhibition of dioxygen binding.

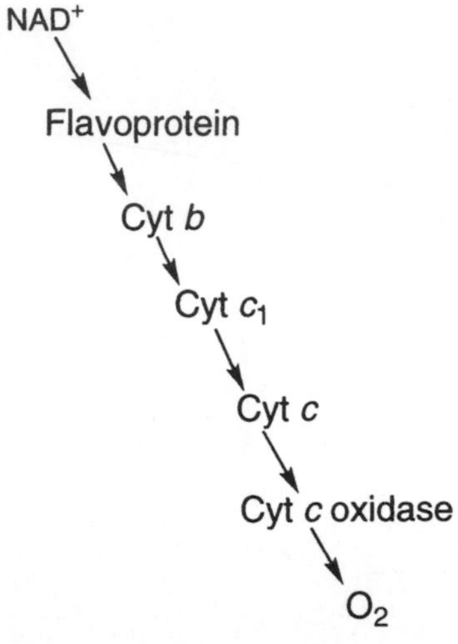

NAD$^+$

Flavoprotein

Cyt *b*

Cyt c_1

Cyt *c*

Cyt *c* oxidase

O$_2$

Fig. 4.9. Part of the mitochondrial electron-transfer chain involving cytochrome *c*.

Cytochrome *c* oxidase

The electron-transfer proteins present in electron-transfer chains are usually bound to cell membranes; the proteins present can either span the phospholipid bilayer component of the membrane or be embedded on one side only. Electrons enter the chain at a potential appropriate to the chemical reaction from which they derive and then run downhill thermodynamically to the reducible electron acceptor via a sequence of redox centres. These centres have small differences (~80mV) in redox potential allowing chemical reversibility.

Cytochrome *c* oxidase is the terminal member of the electron-transfer chain in mitochondrial respiration. It has four active metal ions, two copper and two haem irons, and catalytically reduces dioxygen to water in four steps.

$$O_2 + 4H^+ \rightarrow 2H_2O$$

One copper atom (Cu$_A$) and one haem unit (cytochrome *a*) are involved in electron-transfer processes, whilst the second copper atom (Cu$_B$) and the second haem (cytochrome a_3) carry the dioxygen through the four reduction steps to water. At the same time they neutralize four protons and pump four protons across the membrane in which the enzyme sits (Fig. 4.10). The way in which the enzyme reacts with dioxygen is not well understood and several theories have been advanced. It is probable that at some stage the dioxygen must coordinate to the sixth coordination site in cytochrome *a* and form a heterodimetallic complex with Cu$_B$; a reaction sequence for the reaction starting from Cu$_B$(I)Fe$_{a3}$(II) is shown in Fig. 4.11.

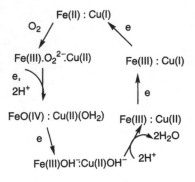

Fig. 4.11. Reaction of cytochrome *c* oxidase (reproduced with permission from Oxford University Press).

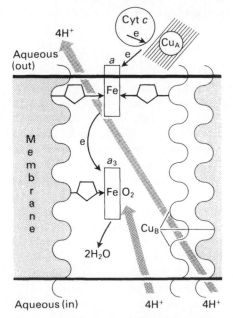

Fig. 4.10. A proposed structural scheme for cytochrome *c* oxidase (reproduced with permission from Oxford University Press).

The immediate environment of the metals is not well characterized. Cytochrome *a* is a low-spin bisimidazole complex and Cu_A has an unusual EPR spectrum with low g values rather like those of a free radical. The dimetallic site is strongly antiferromagnetically coupled, even at room temperature, with an exchange integral in excess of $200cm^{-1}$. EXAFS studies have suggested that chloride can act as a bridging ligand between the iron of cytochrome a_3 and Cu_B in the resting oxidase and so sustain the coupling; the intermetallic separation is *c*. 3.0 Å.

Electron transfer in the respiratory and photosynthetic chains is linked by protons to the synthesis of ATP from ADP and inorganic phosphate. Components of the chain such as cytochrome *c* oxidase act as proton pumps by setting up an electrochemical gradient in response to electron transfer whereby protons are returned across the membrane and down the gradient to drive the synthesis of ATP. This process is described in the chemiosmotic theory of Peter Mitchell.

Peter Mitchell was awarded the Nobel Prize in Chemistry in 1978 for his elucidation of the chemiosmotic theory.

4.4 The photosynthetic pathway

The photosynthetic pathway of green plants may be regarded as an example of an electron-transfer chain. Green plants harvest light energy and convert it to electrical energy through the excitation of chlorophyll. Chlorophyll is a magnesium complex derived from a porphyrin ring in which a double bond in one of the pyrolle rings has been reduced (Fig. 4.12). It is green because it absorbs red and blue light whilst transmitting or reflecting green light. It is

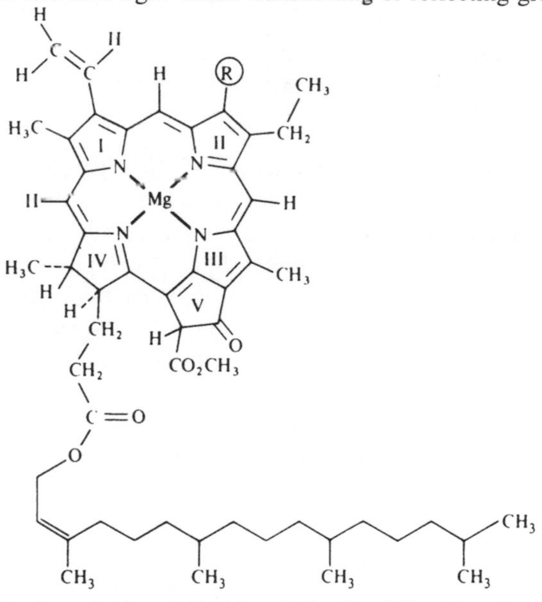

Fig. 4.12. Structure of chlorophyll (chlorophyll *a*, R = CH_3; chlorophyll *b*, R = CHO; the long alkyl chain at the bottom is the phytyl group).

the absorption of photons which excites chlorophylls and the 'excitons' move down light-harvesting chlorophyll protein complexes, or antennae, to the reaction centres Photosystem I (PSI) and Photosystem II (PSII). The two centres act in concert and this is descibed in the 'Z' scheme in Fig. 4.13.

The primary electron donors of these systems are P-700 and P-680 respectively. These are special forms of chlorophyll *a* which have absorption maxima at 700 and 680nm. The PSII centre provides a strong oxidizing agent

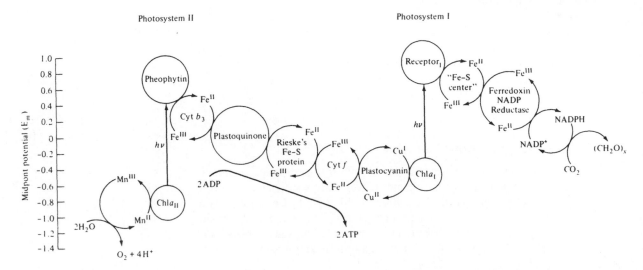

Fig. 4.13. Electron flow in photosystems I and II ('Z'-scheme). The vertical axis gives mid-point redox potential with reducing species (top) and oxidizing species (bottom). Reproduced with permission from Harper Collins.

Harmut Michel, Johann Diesenhofe, and Robert Huber were awarded the 1988 Nobel Prize in Chemistry for their elucidation, via X-ray crystallography, of the structure of the membrane-bound photosynthetic reaction centre of *Rhodopseudomonas viridis*.

which is responsible for the production of dioxygen from water. Upon excitation at this centre, an electron is passed uphill to the primary electron acceptor (pheophytin) and then embarks on a thermodynamically 'downhill' run to PSI via transfer from cytochrome b_3 to plastoquinone, a Rieske 2Fe–2S* protein, cytochrome f, and plastocyanin. At P-700 further excitation occurs to the secondary electron centre from which again runs downhill via an Fe–S centre and ferredoxin NADP reductase to form NADPH which serves to reduce carbon dioxide to carbohydrates. The system can be thought of as consisting of the two PS centres linked by a bridge (or chain) of electron carriers. The electron flow is accompanied by the synthesis of ATP from ADP.

Manganese and PSII

Involvement at the water oxidation–dioxygen evolution centre of PSII is the most important role so far defined for manganese in biological systems. A manganese protein is the site of the reaction

$$2H_2O \rightarrow O_2 + 4H^+ + 4e^-$$

There is no crystallographically defined information for the metal site but it has been established that four manganese atoms are required for activity.

EXAFS studies has indicated two Mn$\cdots$Mn separations of 2.7 and 3.3Å and that the metals are bridged by oxides or hydroxides, possibly in a manner related to haemerythrins. The remaining ligands are oxygen and/or nitrogen donors from the protein amino acids. The tetranuclear aggregate can adopt various oxidation levels during the reaction. These are called the S_n states where $n = 0 - 4$; S_4 is unstable and reverts to S_0 with evolution of dioxygen.

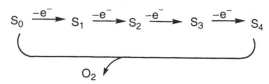

Four one-electron oxidations of the manganese aggregate occur and there is also a requirement for calcium and chloride ions. Although without structural information it is not possible to describe the site in detail it is possible to show a speculative model for it which incorporates the above features (Fig. 4.14).

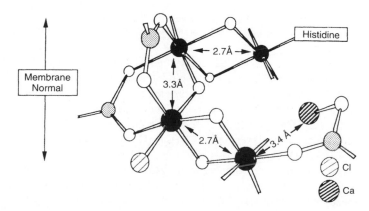

Fig. 4.14. A structural model for the photosynthetic oxygen evolving manganese cluster (reproduced with permission from *Angewandte Chemie*).

The steps in the reaction can be followed by a variety of spectroscopic techniques and from such studies a chemical scheme for dioxygen production at the manganese centre may be presented (Fig. 4.15).

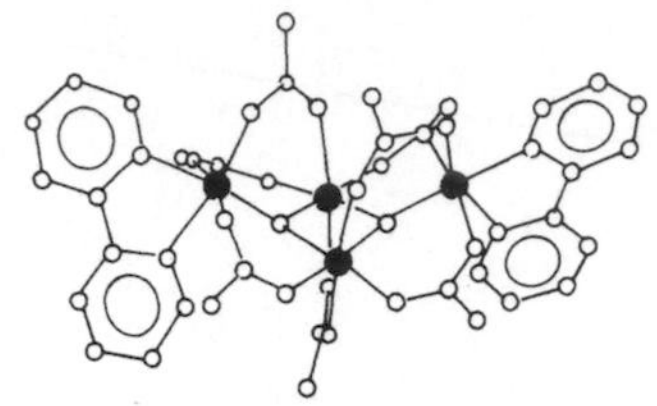

4.3

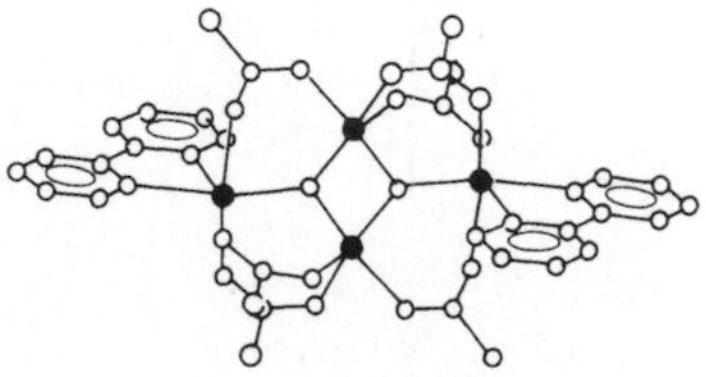

4.4

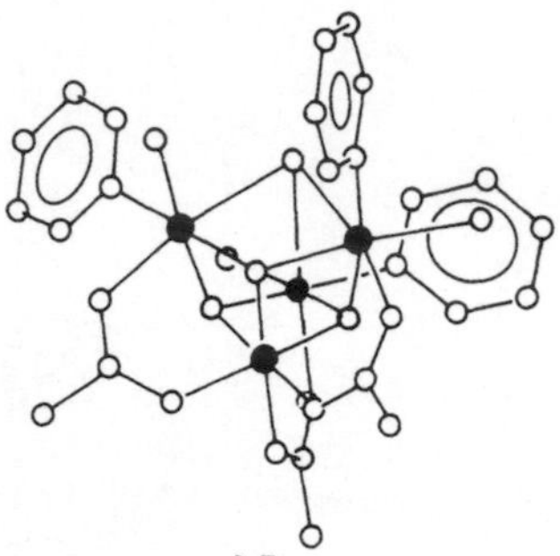

4.5

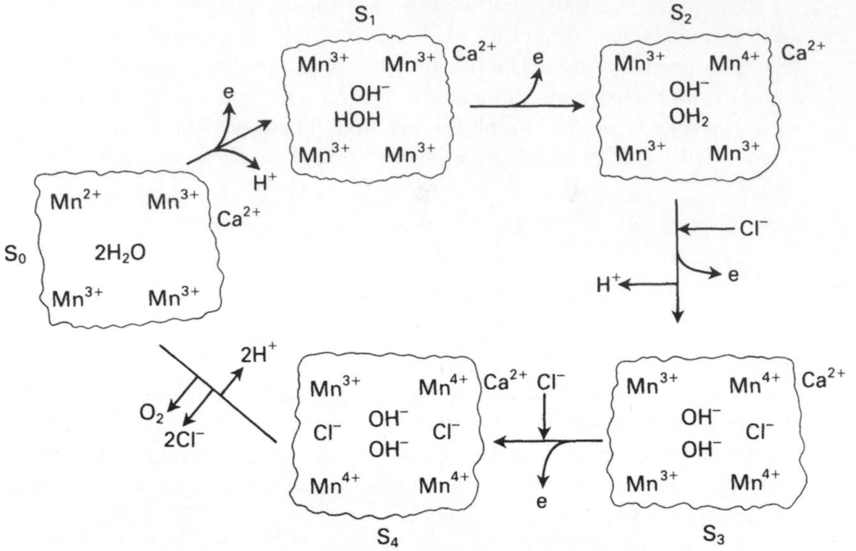

Fig. 4.15. A chemical scheme for O_2 production at a four-manganese cluster involving also Ca^{2+} and Cl^- (reproduced with permission from Oxford University Press).

The possibility that a tetranuclear cluster lies at the active site of PSII has stimulated synthetic chemists to develop model complexes having this nuclearity. Such complexes should feature the Mn...Mn separations idicated by EXAFS (2.7 and 3.3Å) as well as incorporate different manganese oxidation states. George Christou and his colleagues have achieved this using the model complexes $[Mn_4O_2(CH_3CO_2)_7(bipy)_2]^+$ (**4.3**) and $[Mn_4O_2(CH_3CO_2)_6(bipy)_2]$ {bipy = 2,2'-bipyridine} (**4.4**). The former is and unsymmetric complex with a butterfly structure in which exists two different Mn...Mn distances of 2.85 and 3.34(average)Å; but all four manganese are as Mn(III). The latter also has two different Mn...Mn distances (2.78 and 3.84(average)Å] but now has two Mn(II) and two Mn(III) atoms. Subsequently models such as $Mn_3O_3Cl_4(CH_3CO_2)_3(py)_2$ {py = pyridine} (**4.5**) were derived which incorporated chloride anions, two distinct Mn...Mn separations (2.81 and 3.27Å) and a Mn(IV)Mn(III)$_3$ cluster thus modelling the S_2 state of PSII. The development of these model complexes has provided a rapid accompanying development of the chemistry of manganese carboxylates.

5 Dioxygen management—
involvement in enzymes

Enzymes are large proteins which catalyse biochemical reactions often increasing the rate of reaction by factors as high as 10^{12} over the corresponding model reaction; like all proteins they are essentially polypeptides. They have very large molecular weights and the substrate for the enzymatic reaction is generally small and so can only interact with a small portion of the enzyme. This is termed the active site and is often in the form of a cleft, or pocket, in the enzyme structure. A metalloenzyme is an enzyme in which the metal ion serves as the active centre. Enzymes are named after the reaction that they perform—ATP-ase hydrolyses ATP, a hydrolase carries out acid-catalysed hydrolysis, an oxidase is involved in redox reactions.

One requirement for life is the maintenance of molecules in a reduced state even though they are exposed to an oxidizing atmosphere. A further requirement is the production of energy which is accomplished by respiration in which dioxygen is reduced to water. This can be catalysed by cytochrome c oxidase as discussed in the preceding chapter. Dioxygen can be reduced via a series of one-electron transfers (Fig. 5.1). The superoxide and peroxide anions are powerful oxidizing agents and present a threat to life; therefore it is necessary to remove them. Two main methods are available for removing hydrogen peroxide: disproportionation,

Disproportionation is a reaction involving simultaneous reduction and oxidation of the same compound as in

$$2Cu^I Cl \rightarrow Cu^0 + Cu^{II}Cl_2$$

$$2H_2O_2 \rightarrow 2H_2O + O_2$$

and reduction,

$$H_2O_2 + 2H^+ \rightarrow 2H_2O.$$

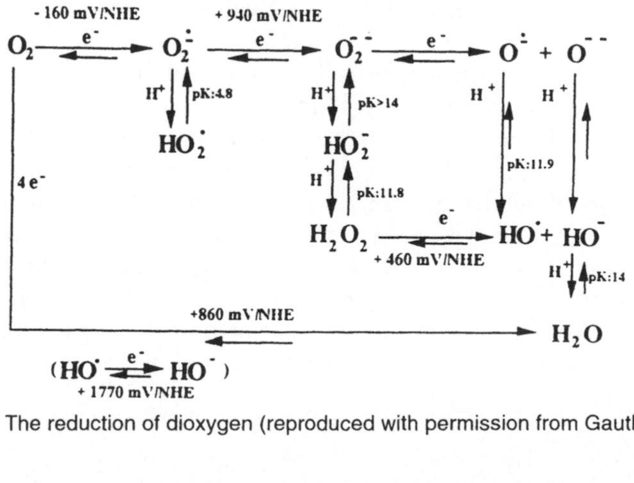

Fig. 5.1. The reduction of dioxygen (reproduced with permission from Gauthier-Vilas).

The former is catalysed by catalases and the latter by peroxidases—both species are haemprotein enzymes. Superoxide dismutases (SODs) catalyse the following reaction

$$2O_2^- + 2H^+ \rightarrow H_2O_2 + O_2$$

with the hydrogen peroxide produced then being removed by catalase.

5.1 Superoxide dismutases

Superoxide dismutases exist either as mononuclear or dinuclear metallo-enzymes. The X-ray crystal structures of mononuclear iron (ex-*Pseudomonas ovalis*) and manganese (ex-*Bacillus thearothermophilus*) SOD and of the heterodinuclear copper–zinc BESOD have been solved.

The structures of the coordination sites in the mononuclear compounds are similar and show the metal to be four-coordinated to three histidine residues and one tyrosine residue (Fig. 5.2). The geometries can be described as distorted trigonal bipyramids with one apical site left empty (an entatic state?). There are no available water molecules and in the iron enzyme the next nearest iron atom is 18Å distant. In contrast in the Mn SOD isolated

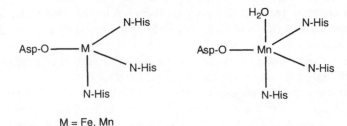

M = Fe, Mn

Fig. 5.2. Schematic representations of the sites in mononuclear SODs.

from *Thermus thermophilus*, there is a water molecule present in the apical site.

The mechanism of action of the mononuclear SODs is not clear but for the iron system the following sequence could occur,

$$O_2^- + LFe(III) \rightarrow LFe(II) + O_2$$
$$O_2^- + LFe(II) + 2H^+ \rightarrow LFe(III) + H_2O_2$$

The structure of Cu–Zn BESOD (Fig. 5.3) shows that the coordination at the copper(II) is distorted square pyramidal with three histidines, a water molecule and a bridging imidazolate residue from His-61 which is shared with the zinc. The zinc is tetrahedral, the coordination being completed by an additional two histidines and an aspartate residue. The copper(II)–zinc separation is 5.4Å. The copper ion is the catalytic centre and the zinc ion has an ancillary structural role. This can be shown by removing the metals to obtain the apoprotein and then reconstituting it with different metals. Every derivative which retains copper(II) in the native site retains almost full activity and no derivative in which this copper has been replaced is active.

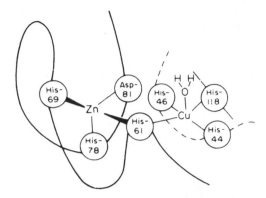

Fig. 5.3. The heterobimetallic site in Cu–Zn BESOD (reproduced with permission from Elsevier).

A proposed mechanism for the enzymatic activity of BESOD (Fig. 5.4) requires that the superoxide replaces the water molecule on the copper(II). The bound superoxide reduces the copper(II) to copper(I) with simultaneous cleavage of the bond between the copper and His-61. Dioxygen is released and a second superoxide now binds to the copper(I) positioning one O atom to form a hydrogen bond with the protonated imidazole of His-61. Electron transfer from the copper(I) coupled with proton transfer from the His-61 gives copper–hydroperoxide which picks up a second hydrogen from the active site water to release hydrogen peroxide.

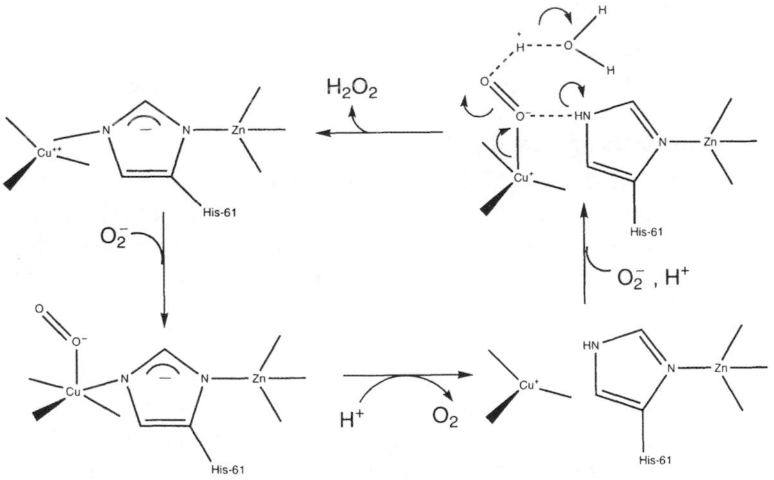

Fig. 5.4. A proposed mechanism for the action of BESOD.

An interesting observation is that the iron and manganese SODs are present in prokaryotes (organisms in which there are no defined nuclei) and the Cu–Zn BESOD is present in eukaryotes (organisms in which the nuclei are defined). This is an illustration of a switch away from a risk-carrying iron enzyme to a safe copper enzyme.

A range of imidazolate-bridged dicopper(II) complexes have been synthesized by Stephen Lippard and his co-workers. The dicopper centre was chosen for modelling as Cu_2BESOD has the same activity as Cu–ZnBESOD. The value of the synthetic complexes as models is shown in their magnetic properties The two copper(II) atoms in the homodinuclear enzyme are antiferromagnetically coupled (J = $-26cm^{-1}$) and in the models this value varies from 0 to $-87cm^{-1}$. For the examples shown, $[Cu(pip)]_2(imid)^{3+}$ and $[(TMPT)_2Cu_2(imid)(ClO_4)_2]^+$ the values are -26.9 and $-25.8cm^{-1}$ respectively. An effective heterodinuclear Cu–Zn model has not yet been synthesized.

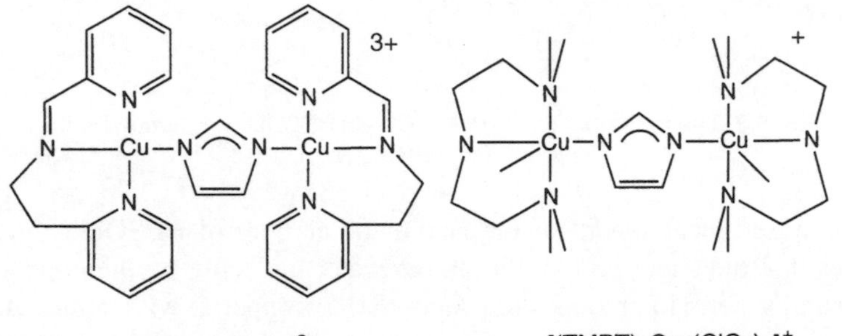

$[Cu(pip)]_2(imid)^{3+}$ $[(TMPT)_2Cu_2(ClO_4)_2]^+$

5.2 Peroxidases and catalases

Peroxidases and catalases are used in the removal of peroxide; they have a strong similarity not only in their function but also in the mechanism of action.

$$H_2O_2 + SubH_2 \rightarrow 2H_2O + Sub \qquad \text{peroxidase}$$
$$H_2O_2 + H_2O_2 \rightarrow 2H_2O + O_2 \qquad \text{catalase}$$

Haem peroxidases and catalases

Horse radish peroxidase, (HRP), present in the roots of horse radish, is perhaps the best studied of these metalloenzymes. The active site is an iron(III) atom from protoporphyrin IX. The fifth coordination site is occupied by a nitrogen atom from histidine and the sixth site is either vacant or contains a water molecule. The iron(III) is high spin at low pH and low spin at high pH.

Upon reaction with hydrogen peroxide the iron undergoes a two-electron oxidation to give green HRP-I which then is reduced in a one-electron step to the red HRP-II. It has been difficult to unravel the mechanism of HRP as collective spectroscopic evidence indicates that both HRP-I and HRP-II contain low-spin oxoiron(IV) species which are virtually unknown in any other chemistry of iron. It has also been found that HRP-I has present a porphyrin radical cation. A scheme for the raction of peroxidase is depicted in Fig. 5.5. In addition to the components ot the above scheme there is a third species HRP-III [P-Fe(II)O$_2$] which is an analogue of oxymyoglobin produced when the peroxidase is treated with a large excess of hydrogen peroxide.

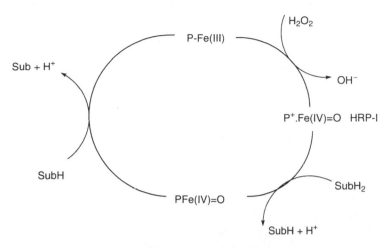

Fig. 5.5. A scheme for the reaction of peroxidase.

The crystal structure of yeast cytochrome c peroxidase (CCP) has been solved and shows that the axial sites are indeed occupied by histidine and water. CCP reacts with peroxides to cleave the O–O bond and generate Compound 1 which is analogous to HRP-1. Compound 1 is then reduced in two successive steps by cytochrome c. As in HRP the intermediates involve iron(IV) but electron abstraction occurs from the amino acid side chain rather than the porphyrin of the heme. The structure of Compound 1 is also reported and in this the electron density pattern close to the iron confirms the suggestion from EXAFS data that there is a strong Fe(IV)=O interaction ($\sim$ 1.67Å)

Another type of haem-iron peroxidase is chloroperoxidase which catalyses the oxidation of chloride ions in the halogenation of a range of substrates again utilizing an oxoiron(IV) intermediate.

$$Cl^- + H_2O_2 \rightarrow ClO^- + H_2O$$

The crystal structure of beef liver catalase shows that the iron(III) atom is five-coordinate with a tyrosine residue occupying one axial position and the second axial position vacant. Despite the structural detail available the mechanism of action of catalase is unclear.

It is interesting to conclude this section with reference to a recent lecture by Max Perutz in which he reminisces as follows: 'At about the time that Pauling delivered his discourse [at the Royal Institution in 1948] my mentor David Keilin asked me: "Haemoglobin, peroxidase and catalase all contain the same haem. What gives them their very different properties?" Keilin's tone implied that as a structural chemist I ought to know the answer, but I was clueless. Had I more imagination, I ought to have replied that these different properties were likely to be due to the different electrostatic effects exercised on the haem by the different proteins.' In all haem proteins the haem iron is attached to the protein by an amino acid residue—histidine in haemoglobin and peroxidase, tyrosine in catalase, cysteine in cytochrome

P–450 and so it is intriguing to think that the different properties are related to the different electron densities at the metal and to the different electrostatic effects induced by these different protein environments.

Vanadium bromoperoxidase

Vanadium bromoperoxidases (V-BrPA) have been isolated from marine brown and red algae. They catalyse the oxidation of bromide by hydrogen peroxide and the concurrent formation of carbon–halogen bonds, or, in the absence of an organic substrate, the generation of singlet dioxygen from the bromide-assisted disproportionation of hydrogen peroxide. In contrast to the corresponding haem iron haloperoxidases, the production of dioxygen only occurs in the presence of bromide or iodide anions. The halide oxidation selectivity of V-BrPO was first thought to be restricted to bromide or iodide but it is now known that V-BrPO can catalyse the oxidation of chlorine by hydrogen peroxide leading to the chlorination of selected organic substrates.

EPR studies have shown that the metal is present as vanadium(V); when the metal is reduced to the vanadium(IV) state the enzyme loses its brominating activity. EXAFS studies suggest that it is in a distorted octahedral environment with two histidine nitrogen atoms and one oxo group bound to the metal; the remaining three sites are probably occupied by three oxygen atoms from undetermined amino acids. A scheme for the reaction carried out has been advanced (Fig. 5.6). However the precise nature of the 'intermediate' is not established and the exact role of the metal is not yet

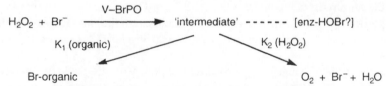

Fig. 5.6. A reaction scheme for vanadium bromoperoxidase.

known. The vanadium could act as an electron-transfer catalyst or a Lewis acid catalyst; it is also possible that it is the peroxide and not the metal that is involved in the reaction.

5.3 Oxidases and oxygenases

Oxidases catalyse electron transfer to dioxygen which is reduced to superoxide, peroxide, or water. Oxygenases catalyse reactions in which the atoms of dioxygen are incorporated into organic substrates. In monooxygenases one oxygen atom is inserted, with the second being reduced to water, and in dioxygenases both oxygen atoms are inserted into the substrate. Dioxygen is kinetically inert and requires activation by binding to a low-valent metal centre, usually iron(II) or copper(I). Control over the transfer of electron density from the metal to dioxygen dictates the biological role of the metalloprotein or metalloenzyme, determining whether it will function as an oxygen carrier, an oxidase, or in oxygenase hydroxylation reactions.

Oxidases

The haem oxidase cytochrome *c* oxidase and the blue copper oxidases laccase and ascorbate oxidase have been referred to in Chapter 4. All are involved in the four-electron reduction of dioxygen to water and the copper enzymes concomitantly oxidize a substrate. Ascorbate oxidase oxidizes L-ascorbate to dehydroascorbate, and laccase is less specific and oxidizes a range of substrates but particularly *p*-phenols to quinones.

There are other non-blue copper oxidases. Amine oxidases catalyse the oxidative deamination of biological amines generating aldehydes, hydrogen peroxide, and ammonia.

$$RCH_2NH_2 + O_2 \rightarrow RCHO + H_2O_2 + NH_3$$

The copper site has been shown by EPR to contain at least two nitrogen atoms, probably from histidines, but it is not certain that the copper actually activates dioxygen. Galactose oxidase catalyses the oxidation of primary alcohols to the corresponding aldehyde with reduction of dioxygen to hydrogen peroxide. The copper site (Fig. 5.7), as revealed by X-ray crystallography by Simon Phillips and his co-workers, is unusual and quite dissimilar from that found for the oxidative copper enzymes ascorbate oxidase and superoxide dismutase. The copper(II) atom (Type II) is coordinated by two equatorial histidines (at 2.11 and 2.15Å), two tyrosines (one axial at 2.69Å and one equatorial at 1.94Å), and an acetate anion (at 2.27Å) in a distorted square pyramidal environment. The acetate anion (which arises from use of an acetate buffer at pH 7.0) can be replaced by a water molecule at pH 7.0 which is the pH region at which the oxidase is active. This water molecule is *c.* 2.8Å from the copper and so the very distorted geometry could induce an entatic situation leading to a facile redox shuttle between copper(II) and copper(I) after addition of the substrate. Furthermore D-galactose, the substrate, could replace the water and bind to the equatorial site.

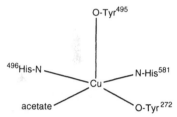

Fig. 5.7. The copper site in galactose oxidase.

Monooxygenases

Monooxygenase activity can occur at mono- or dinuclear metal sites; these are represented here by cythrome P-450, a mononuclear haem enzyme, tyrosinase, a dinuclear copper enzyme, and methane monooxygenase, a dinuclear iron non-haem enzyme.

Cytochromes P-450

These are haem enzymes which act as mono-oxygenases and use dioxygen to catalyse aromatic and aliphatic hydroxylation reactions and various other oxidation reactions. The name of the enzymes derives from the fact that their carbon monoxide adducts have absorption bands at 450nm. Their reaction requires a hydrogen donor such as NADH or a flavin, below denoted by AH_2.

Cytochrome P-450 consists of a haem group with the iron coordinated to the protein by a cysteine sulphur group. This has been shown through the crystal structure of the camphor bound form of cytochrome P-450 from

Pseudomonas putida. In the resting state the iron is low spin and there should be an additional ligand, probably water, in the sixth coordination site which is then lost to form the active five coordinate high-spin iron(III) compound which reacts with the substrate. The iron(III) is then reduced to iron(II) and once in this state it can bind dioxygen to give a diamagnetic oxycomplex similar to oxymyoglobin. Addition of a second electron to the super-oxocomplex is followed by heterolytic cleavage of the O–O(peroxo) bond. One oxygen is lost as water and the other forms a perferryl intermediate which is involved in a two-electron oxidation of the substrate to give the product and regeneration of the iron(II) state of the enzyme (Fig. 5.8).

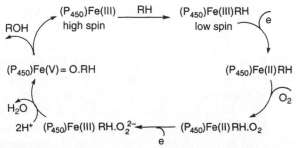

Fig. 5.8. A possible catalytical pathway for cytochrome P-450.

The difficult area of this scheme is in the mechanism of cleavage of the O–O bond. The first product of cleavage could be the perferryl complex shown in the scheme or it could be a porphyrin radical cation as in $[Fe(IV)(O^{2-})P^{\cdot}]^{+}$. The difficulty of elucidating the mechanism is related to the paucity of Fe(IV) and Fe(V) chemistry in systems other than the haem proteins. Studies on the hydroxylation of alkanes by iodosylbenzene and model iron porphyrins has confirmed the plausibility of a perferryl intermediate; structural models have also been presented for the active iron (III) site (Fig. 5.9).

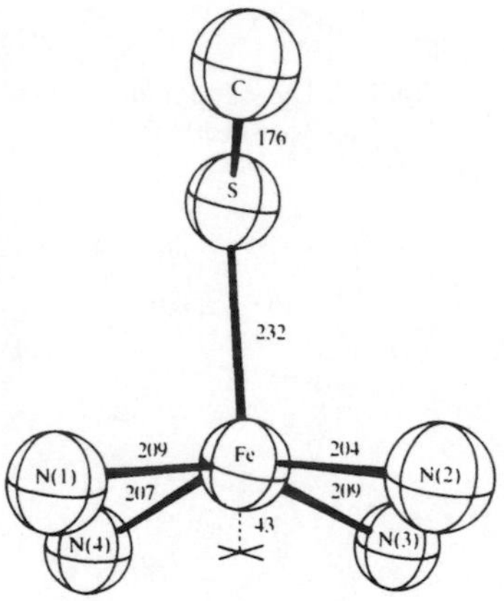

Fig. 5.9. Molecular structure of a model for the active site in cytochrome P-450; the basal ligand is the dimethyl ester of proto-porphyrin IX and the apical ligand is 4-nitrothiophenolate (bond distances in picometers) (reproduced with permission from the American Chemical Society).

Tyrosinase

Tyrosinase is widely found in microorganisms, plants, and animals where it acts as a monooxygenase catalysing the *o*-hydroxylation of monophenols to *o*-diphenols, and as a two-electron oxidase catalysing the oxidation of *o*-diphenols to *o*-quinones. The former behaviour is termed cresolase activity and the latter is catecholase activity. Chemical and spectroscopic evidence indicates that the tyrosinases have a Type III coupled dinuclear active site closely similar to that found in the haemocyanins.

A mechanism for the catalytic activity of tyrosinase has been advanced by Edward Solomon and his co-workers (Fig. 5.10) Deoxytyrosinase reacts with dioxygen to give oxytyrosinase; the phenolic substrate then coordinates axially to one of the copper atoms in the oxytyrosinase. There is then a rearangement through a five-coordinate intermediate accompanied by *ortho*-hydroxylation of the phenol, loss of water, and coordination of the diphenolic product. Intramolecular electron transfer leads to the *ortho*-benzoquinone product and regenerates deoxytyrosinase.

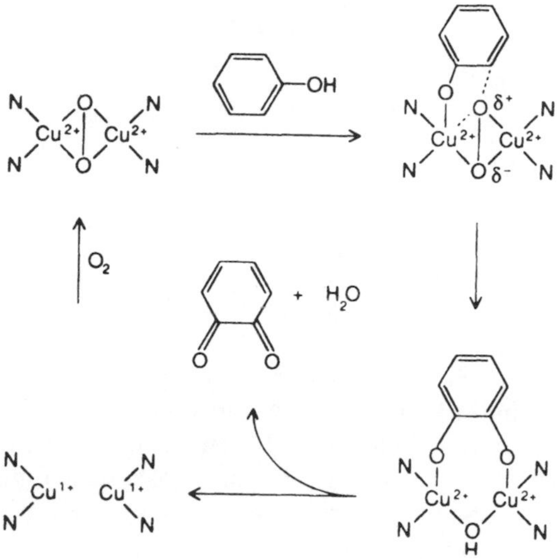

Fig. 5.10. A mechanistic proposal for the reaction of tyrosinase.

Methane monooxygenase

The enzyme methane monooxygenase (MMO) catalyses the conversion of methane to methanol in methanotropic bacteria for which methane serves as the sole source of energy and carbon.

$$CH_4 + O_2 + NADH + H^+ \rightarrow CH_3OH + H_2O + NAD$$

There are two kinds of MMO, a membrane-bound enzyme containing copper and a water soluble enzyme containing non-haem iron. EPR studies on the hydroxylase component of MMO from *Methylococcus capsulatus (Bath)* indicated that it contains a diiron centre which was at first thought to be of a similar type to those found in hemerythrins and ribonucleotide reductase B2. However a detailed study of the hydroxylase components, involving EXAFS, mass spectrometry, and magnetochemistry, indicated that there were striking differences between the diiron sites in these systems and that MMO did not contain a Fe–O–Fe bridge. A novel type of Fe–Fe bridge was indicated containing an alkoxo-, hydroxo-, or monodentate carboxylato bridge, together with one, or two, bridging carboxylates.

The X-ray crystal structure of the MMO hydroxylase from *M. capsulatus (Bath)*, solved by Stephen Lippard and his coworkers, revealed that the geometry of the diiron core has the two pseudo-octahedral iron atoms separated by 3.4Å and linked by two non-protein bridging ligands (Fig. 5.11).

A methanotroph is an organism which grows with methane as its sole source of carbon and energy. Methanotrophic bacteria play an important role in the carbon cycle as they limit the release of methane, a 'greenhouse gas', into the atmosphere.

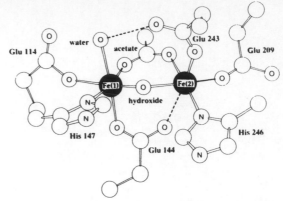

Fig. 5.11. Schematic representation of the dinuclear site in MMO hydroxylase (reproduced with permission from *Nature*).

One is a hydroxide anion and the second is an acetate anion. Ammonium acetate was present in the buffer solution from which the crystal was grown and so this ligand may not be coordinated to the dinuclear site in the cell. It was suggested that the space occupied by the acetate ligand might indicate where the oxidized substrate, the methoxide anion, would be prior to its protonation and release from the active site.

A mechanism for the catalytic activity has been presented by Howard Dalton which assumes the presence of a hydroxobridged diiron(III) unit (**5.1**) (Fig. 5.12). Electrons are fed into the unit and the active diiron(II) unit (**5.2**) then adds dioxygen to give a peroxide intermediate with a stabilizing

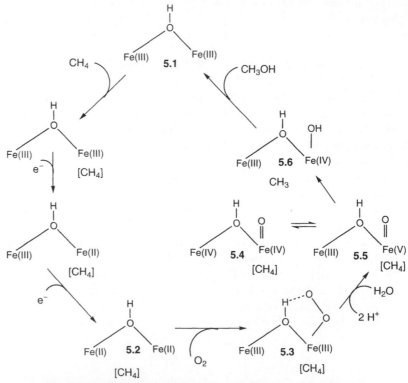

Fig. 5.12. A mechanism for the activity of methane monooxygenase .

hydrogen bond from the hydroxo-bridge (**5.3**). Proton donation then occurs to the outer oxygen of the peroxo-ligand leading to the formation of water and a ferryl unit which may be written in either Fe(IV)–Fe(IV)O (**5.4**) or Fe(III)–Fe(V)O (**5.5**) form. This can then accept a hydrogen from the methane leading on to generation of methanol via donation of –OH from the Fe(IV)–OH unit (**5.6**) and regeneration of the original diiron(III) unit.

Dioxygenases

The metal-containing dioxygenases usually contain iron in a non-haem environment. An example of their activity is the oxidative cleavage of 1,2-catechols by an extra-diol (a) or intradiol (b) pathway. The pattern of the reaction can be monitored by ^{18}O-labelling experiments. Catechol-1,2-dioxygenase catalyses the intradiol cleavage of catechol by oxygen to give *cis,cis*-muconic acid, so incorporating both of the oxygen atoms. EXAFS studies have suggested that the iron(III) is six-coordinate having two histidine and two *cis*-tyrosine ligands together with a water molecule and a sixth ligand which could be easily dissociated. The catechol binds to the enzyme before oxidation and so is coordinated, probably monodentate, to the iron.

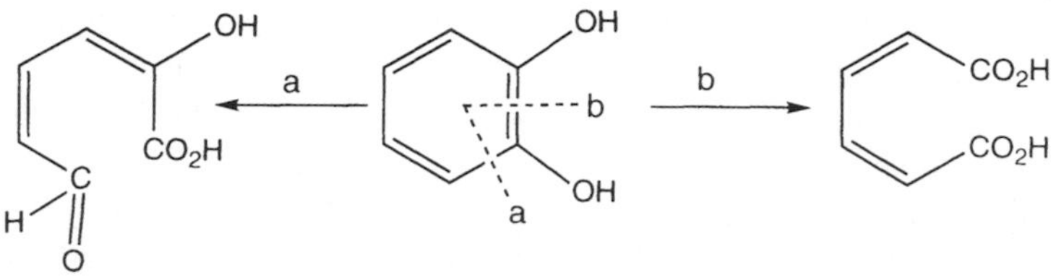

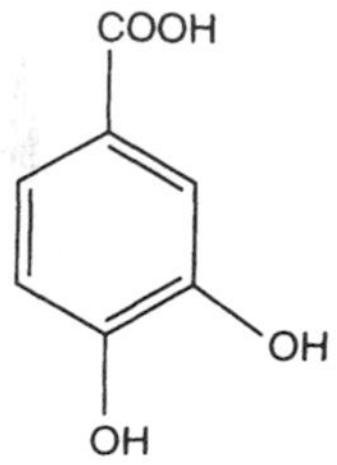

protocatechuic acid

Catecholate

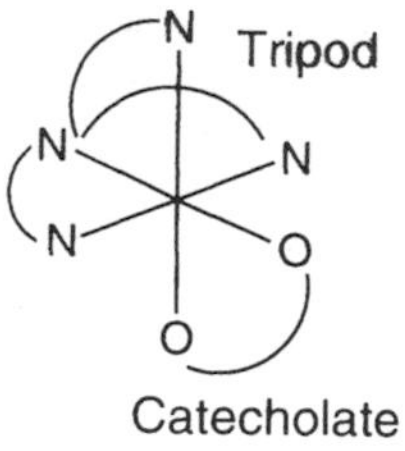

Fig. 5.13. Schematic diagram of the [Fe(III) (TPA)DBC]$^+$ cation.

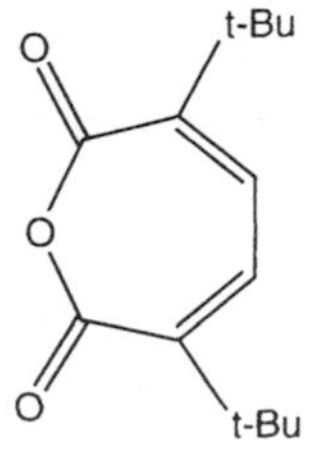

Catechol-2,3-dioxygenase is an example of an extradiol dioxygenase; it is colourless and EPR-silent suggesting the presence of iron(II). The Mössbauer spectrum confirms this and is unaffected by substrate binding, thus raising the question of whether the substrate actually binds to the metal.

The crystal structure of protocatechate-3,4-dioxygenase, which catalyses the intradiol cleavage of protocatechuic acid to β-carboxy-*cis,cis*-muconic acid, has been solved and shows that the iron(III) is in a distorted trigonal bipyramidal environment. The iron is bonded by two tyrosines and two histidines with a solvent molecule in the fifth site. One tyrosine and one histidine occupy axial positions. A highly reactive functional model for catechol dioxygenase activity has been constructed by Lawrence Que and his co-workers, [Fe(III) (TPA)DBC]BPh$_4$, [TPA = *tris*-(2-pyridylmethyl)amine

5.7

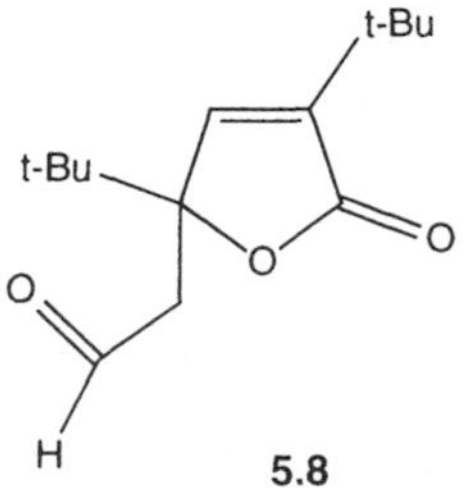

5.8

and DBC = 3,5-*t*-butylcatecholate] (Fig. 5.13). The structure of this complex reveals a distorted octahedral iron(III) centre with long iron–catechol oxygen bonds. Reaction of this complex with dioxygen facilitates oxidative cleavage of the catechol to give compounds (**5.7**) and (**5.8**).

The role of the enzyme indole-2,3-dioxygenase serves to illustrate the wider application of dioxygenases as this enzyme from the small intestine of the rabbit causes cleavage of the indole ring in the amino acid tryptophan and other related indoleamines.

indoleamine   anthranisoylalkyamine

5.4 Ribonucleotide reductase

Ribonucleotide reductase (RR) catalyses the conversion of ribonucleotides to deoxyribonucleotide di- or triphosphates, the first step in DNA synthesis. *E. coli* RR contains two subunits, a and b, and the two dimeric forms a_2 and b_2 are known as B_1 and B_2 respectively. The a units contain the substrate binding sites but both B_1 and B_2 contribute to the active site of the enzyme. RRB2 contains a stable free radical—it was the first example to be found in proteins—and EPR was used to characterize this as a protein-derived tyrosyl radical located in the vicinity of a dinuclear iron centre (galactose oxidase also has a tyrosine radical close to the active site). The reaction catalysed by RR has been proposed to proceed via a free-radical mechanism involving the cyclic transfer of the pre-existing radical to the substrate and back. The diiron centre in RRB2 showed spectral properties closely similar to those found in Met-haemerythrin and this led to the proposal that, by analogy, the site was a similarly constructed bioctahedral site. The X-ray crystal structure of the oxidized form of *E. coli* RRB2 (Fig. 5.14) confirmed that the Fe–O–Fe bridge was present, as in Met-haemerythrin, and showed that there are two dinuclear iron(III) centres, separated by ~25Å, in the dimeric molecule. The dinuclear centre has an oxygen-dominated first ligand sphere and the iron atoms, separated by 3.3Å, are, in contrast to Met-haemerythrin, doubly bridged by a single glutamate carboxylate residue and the oxo-ligand.

The coordination geometries of the iron atoms are quite different from those in Met-haemerythrin and from each other. One iron atom is terminally coordinated by two monovalent glutamates, a histidine, and a water molecule in a regular octahedral geometry, whereas the second iron atom is coordinated further by a bidentate aspartate, a water molecule, and a histidine. The geometry has both octahedral and trigonal bipyramidal features because of the chelating aspartate which can be considered as occupying two

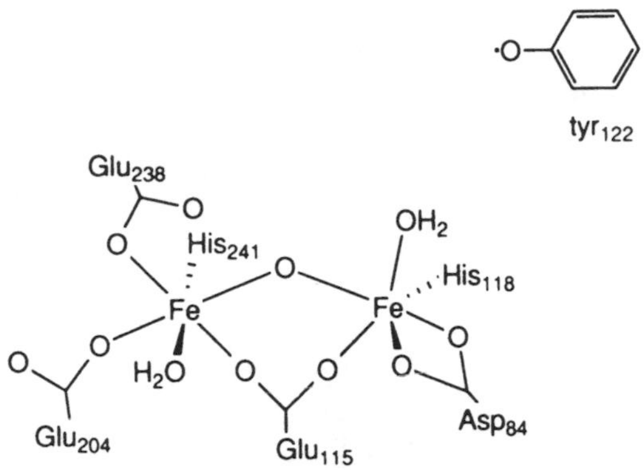

Fig. 5.14. Schematic of the diiron(III) site in RRB2 from *E. coli*; the relative position of the tyrosine radical that is 5Å from the diiron site is shown (reproduced with permission from John Wiley).

donor sites and so giving a severely distorted octahedron,or as utilizing one equatorial site of a trigonal bipyramid. Three redox states are possible for the dinuclear iron centre. The diiron(II) and diiron(III) have been well characterized but it is only recently that the mixed-valence Fe(II)Fe(III) state has been detected.

Although the crystal structure of the reduced form of RRB2 has not yet been solved the crystal structure of Mn(II)-reconstituted RRB2 (Figure. 5.15) shows that the metals are bridged by two bidentate carboxylates, Glu-115 and Glu-238, and that there is no oxo-bridge. The diiron(II) site in $RRB2_{red}$ is then proposed to be similar and a mechanism for the reduction of the diiron centre in $RRB2_{ox}$ has been presented (Fig. 5.16). This involves protonation of the oxo-bridge and carboxylate shifts.

The dinuclear site in the oxidized form of ribonucleotide reductase clearly presents a challenge to the model-maker, not only in the reconstruction of the double bridge but also in the synthesis of a complex which has present three distinct binding modes (monodentate, bidentate, and bridging) for the four

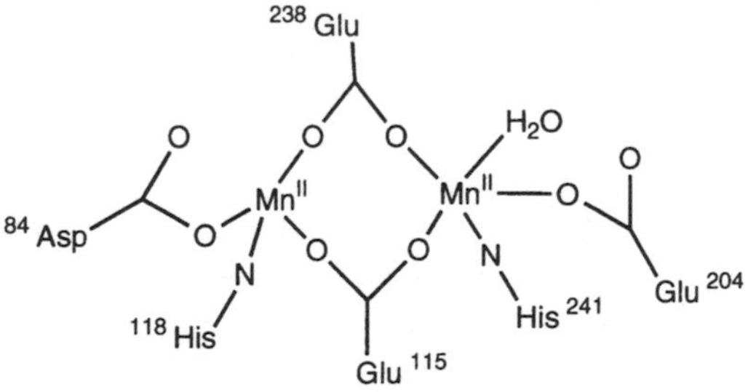

Fig. 5.15. The dinuclear site in manganese(II)-reconstituted RRB2.

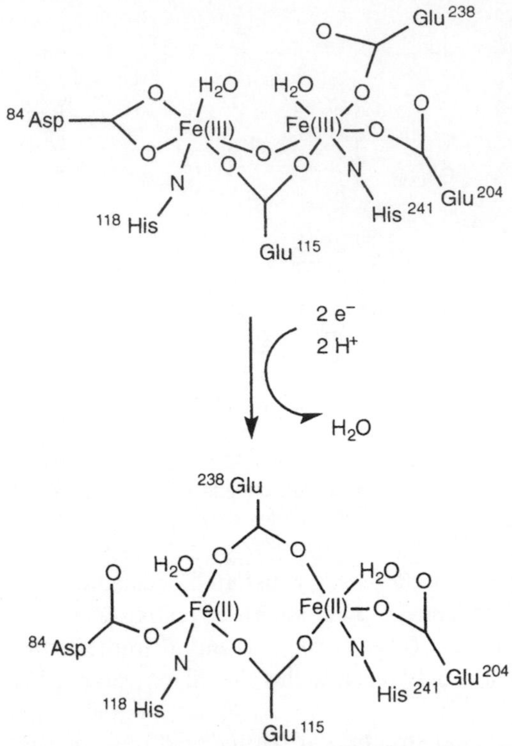

Fig. 5.16. Proposed mechanism for the reduction of the diiron centre in RRB2 (after Fontecave).

accompanying carboxylates. Lawrence Que has presented a bis(μ-carbo-xylato-O,O')diiron(II) complex, $[Fe_2(O_2CCH_3)_2(TPA)_2](BPh_4)_2$(**1**)(TPA = tris(2-pyridyl methyl)amine), (Fig 5.17), which interacts with dioxygen in the presence of protons to give a (μ-oxo)(μ-carboxylato O,O')diiron(III) complex

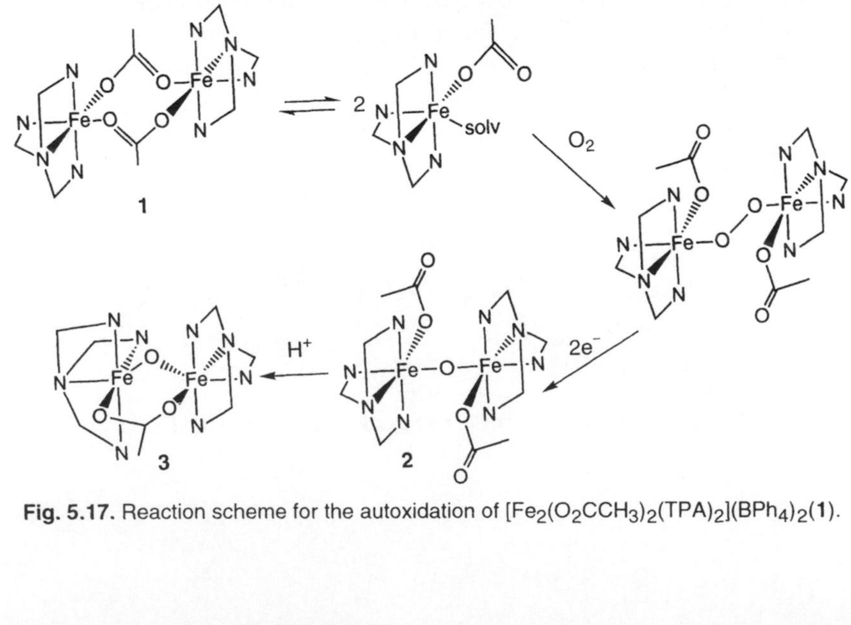

Fig. 5.17. Reaction scheme for the autoxidation of $[Fe_2(O_2CCH_3)_2(TPA)_2](BPh_4)_2$(**1**).

(**3**), and commented that this chemistry might be analogous to the structural changes associated with the diiron site in ribonucleotide reductase.

The relevance of the bis(μ-carboxylato-O,O')diiron(II) centre as a structural motif was suggested by the crystal structure of Mn(II)-reconstituted RRB2. There are differences between the structures of the model and of RRB2. The bridging carboxylates are *syn-anti* in the model with an intermetallic separation of 4.3Å but they are likely to be *syn-syn* in the protein, as they are in the manganese-reconstituted protein and as suggested by the shorter intermetallic separaton of 3.3Å; the nitrogen and oxygen content of the donor atom sets are also different. Never the less, the model may be regarded as a reactivity model because it reacts with dioxygen to yield a mixed oxo-, carboxylato-bridge in the oxidized dimer (**3**) similar to that found in RRB2$_{ox}$.

The structural relationship between MMO and RRB2 provides an interesting example of how Nature can use closely related structural features to catalyse very different chemical reactions.

6 More metalloenzymes

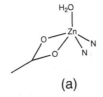

(a)

(b)

(c)

Fig. 6.1. The coordination spheres of the active mononuclear zinc sites in (a) carboxypeptidase A, (b) carbonic anhydrase II, and (c) alcohol dehydrogenase.

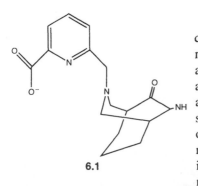

6.1

Several metalloenzymes using iron and copper have been met in the preceding chapters; this chapter is primarily concerned with the enzymatic activity of zinc, cobalt, and molybdenum.

6.1 Zinc, superacid

Zinc is the second most abundant trace element in man and is required as an integral component of over 100 enzymes in different species of all phyla. It can play an active catalytic role, generally as a strong Lewis acid, or it can act in regulatory or structural roles.

The unique feature of all structurally characterized active mononuclear zinc sites (Table 6.1) is a water molecule which can be activated by ionization, by polarization, or be poised for displacement by a substrate (Fig. 6.2).

Table 6.1. Ligands present at active mononuclear zinc sites

Enzyme	Ligand 1	Ligand 2	Ligand 3	Ligand 4
Carbonic anhydrase	His	His	His	H_2O
Carboxypeptidase	His	Glu	His	H_2O
Alcohol dehydrogenase	Cys	His	Cys	H_2O
Thermolysin	His	His	His	H_2O
β-Lactamase	His	His	Glu	H_2O
Bacillus cereus neutral protease	His	His	His	H_2O

Two generalized mechanistic pathways have been suggested for the catalytic action of zinc enzymes (Fig. 6.2). In the first the coordinated water molecule is deprotonated to leave a bound hydroxide which can then either attack the carbon atom of the carbonyl group in a substrate or act as a proton acceptor. Ionization of the activated water molecule (enhancement of the acidity of the coordinated water) promotes the nucleophilicity of the bound solvent molecule; this process can be assisted by polarization by a base form of a proximal amino acid. The pK_a of metal-free water is 15.7 and this can be reduced to around 10 in $[Zn(H_2O)_6]^{2+}$; it can be further reduced to around 7 in the tridentate complex $[(L)Zn(H_2O)]^{2+}$ (L is depicted as **6.1**). In the second mechanism the substrate attaches directly to the zinc through a carbonyl group which is then polarized inducing nucleophilic attack at the carbon atom of the carbonyl group.

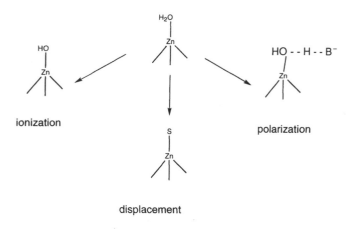

Fig. 6.2. The functions of the H_2O ligand at the active zinc site. S = substrate, B = base.

Zinc has several chemical features which benefit its presence at the catalytic sites. The ready formation of low coordination number sites which have more acidity than high coordination number sites, the easy deformation of the coordination site geometry and the accessible change of coordination number from four to five to six, are all of assistance at a catalytic site where there must be flexibility throughout the catalytic cycle. Zinc can exchange its ligands readily but must not itself be lost from the protein. Zinc, as a d^{10} ion, has no reduction–oxidation chemistry and so it is likely that many zinc enzymes are used in situations where the presence of a redox active metal would promote reactions injurious to the host. Only two examples of zinc enzyme activity will be discussed here.

Carbonic anhydrase II

Human carbonic anhydrase II (HCA) was first recognized as a zinc enzyme in 1940. It is found predominantly in red blood cells where it catalyses, with a high turnover rate (10^6 sec^{-1} at pH 9 and 25°C), the reversible hydration of carbon dioxide to form the bicarbonate anion and a proton via a tandem chemical process.

$$HO^- + CO_2 \rightleftharpoons HCO_3^-$$
$$H_2O \rightleftharpoons OH^- + H^+$$
$$\overline{}$$
$$H_2O + CO_2 \rightleftharpoons HCO_3^- + H^+$$

In the absence of a catalyst, hydration of carbon dioxide occurs relatively slowly and HCA enhances the rate by a factor of 10^7. HCA can also catalyse hydrolysis of esters and aldehydes.

The structure of the metalloenzyme shows that the zinc atom lies near to the bottom of a cleft some 15Å deep. It is ligated to the imidazole nitrogen

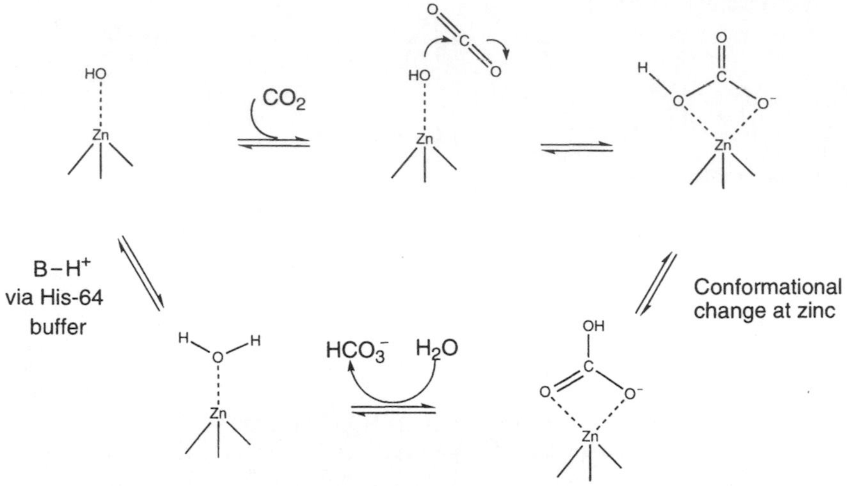

Fig.6.3. A general mechanistic profile for the catalysis of CO_2 hydration.

atoms of three histidine residues and to a water molecule in a distorted tetrahedral array. A general mechanistic profile for the catalysis of CO_2 hydration is presented in Fig 6.3.

The first sequence involves an outer sphere association of the substrate with the enzyme, via nucleophilic attack at the carbon atom of the CO_2, followed by chemical conversion of the substrate to the coordinated product. After a conformational rearrangement of this product at the metal centre the second sequence involves product dissociation and regeneration of the catalytically active species by substitution with water and participation of the neighbouring His-64 as a proton acceptor. While certain structural features such as the function of the pentacoordinate zinc and the degree of CO_2–Zn^{2+} interaction in the enzyme–substrate association have not yet been fully elucidated it is not believed that a precatalytic inner sphere $O=C=O \rightarrow Zn^{2+}$ interaction is operating in HCA.

A functional model for HCA activity has been derived from the highly substituted pyrazolylborate ligand HBL_3^- (L = 3-*tert*-butyl-5-methyl-pyrazolyl; Section 3.2). This ligands caps one face of the tetrahedral zinc leaving an apical site available for further interaction. $[(HBL_3)Zn]OH$ is synthesized and reacts reversibly with CO_2 to give a bicarbonato complex. Hydrolysis of this complex regenerates the hydroxo-complex thus paralleling the facile displacement and regeneration of the hydroxozinc species that is critically involved in the proposed catalytic cycle for HCA.

Carboxypeptidase A

Bovine carboxypeptidase A (CPA) is a metalloexopeptidase which hydrolyses C-terminal amino acids from peptide substrates and has a preference towards those substrates possessing large, hydrophobic C-terminal side chains as does phenylalanine. It was the first zinc enzyme to be characterized crystallographically. The zinc atom was shown to be ligated within a cleft by the imidazole nitrogen atoms of two histidine residues, a

Isomorphous replacement: as a d^{10} ion zinc is said to be spectroscopically silent with regard to techniques such as EPR, UV, and visible spectroscopies the technique of isomorphous replacement can be used to probe the nature of the site in an enzyme in the absence of X-ray crystallographic techniques. If the apo-enzyme, in this example HCA, is reconstituted with other metals, it is possible to use a range of spectroscopic techniques. On reconstitution with Co^{2+} HCA retains about 50% of its activity; Cd^{2+} and Mn^{2+} have only slight activity and Cu^{2+} and Ni^{2+} have negligible activity. This is related to the geometry at the site and as, of the metals used here, Co^{2+} is capable of adopting to the required tetrahedral geometry then it partly restores the enzymatic activity. A similar pattern emerges for carboxypeptidase.

bidentate glutamate residue and a water molecule leading to a distorted pentacoordination geometry.

The structures of enzyme–pseudosubstrate complexes have provided a model for the precatalytic interaction and a mechanistic proposal for the action of CPA is outlined in Fig. 6.4.

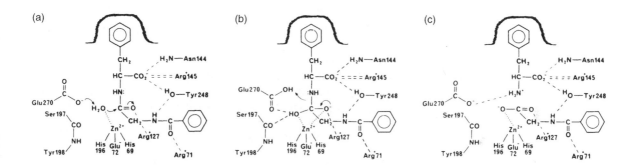

Fig. 6.4. A mechanistic proposal for the action of CPA.(reproduced with permission from the American Society of Chemistry).

A peptide substrate bearing a terminal phenylalanine residue bound to the active site of CPA with the terminal phenyl group residing in the hydrophobic pocket of the enzyme is depicted in Fig. 6.4a. The amino acids Arg-145 and Tyr-248 are hydrogen-bonded to the terminal carboxylate of the substrate and Tyr-248 is also hydrogen-bonded to the amide-NH of the penultimate peptide bond. This interaction might provide a specificity towards substrates having such bonds available. The scissile carbonyl group is polarized by Arg-127 thus encouraging nucleophilic attack by the water molecule promoted by the zinc and assisted by Glu-270. This pathway has been termed a promoted-water reaction.

In Fig. 6.4b the tetrahedral organic intermediate is stabilized by interaction of the *gem*-diol to provide a pentacoordinate zinc. The intermediate can collapse to the product complex Fig. 6.4c via proton donation from Glu-270 prior to product release which may be facilitated by unfavourable steric interactions between Glu-270 and the product carboxylate.

It has been proposed that this mechanistic sequence may be general to many other zinc proteases. It is not, however, absolute and an inner sphere mechanism involving loss of water and direct substrate attachment to zinc has been considered in the catalytic action of thermolysin. The enzymes HCA and CPA share a simple catalytic function in the attack of hydroxide at the substrate C=O bond. The fate of the species formed differentiates between the class of enzyme, as in HCA the nucleophilic adduct is the product, whereas in CPA the nucleophilic adduct is an intermediate.

HCA

nucleophilic adduct is the product

CPA

nucleophilic adduct is a reaction intermediate

6.2 The structural role of zinc

Zinc can play a structural role in certain proteins. Alcohol dehydrogenase has two zinc atoms present one of which provides an active site for the reaction $RCH_2OH + NAD^+ \rightarrow RCHO + NADH + H^+$ (NAD is nicotinamide adenine dinucleotide) with the second zinc atom being tetrahedrally coordinated to four sulphur atoms from cysteine ligands and so being inaccessible to water; the zinc atom in aspartate transcarbamylase is similarly coordinated. The zinc atoms control local protein-folding and conformation, and in so doing can cross-link the protein chain possibly taking the place of a disulphide bridge. The zinc present in insulin is coordinated by three nitrogen atoms from histidines and three water molecules in an irregular octahedral environment which is believed to have a structural function.

Transcription factors regulate gene expression, an essential feature of which is the binding of a regulatory protein to the recognition sequence of the appropriate gene. Many such proteins have been found to have embedded

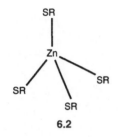

6.2

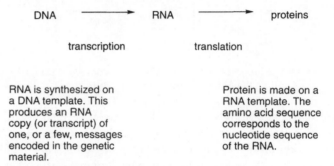

DNA $\longrightarrow$ RNA $\longrightarrow$ proteins

transcription   translation

RNA is synthesized on a DNA template. This produces an RNA copy (or transcript) of one, or a few, messages encoded in the genetic material.

Protein is made on a RNA template. The amino acid sequence corresponds to the nucleotide sequence of the RNA.

in their structure a zinc-containing motif that serves to bind to DNA. The role of the zinc in these DNA-binding proteins appears to be purely structural, and has the advantage over the formation of disulphide links that zinc cannot be reduced in the reducing atmosphere of the cell.

The motif was conveniently called a zinc finger and is a module that can be used either singly, or in tandem, as was originally discovered in the *Xenopus* transcription factor IIIA, to recognize DNA (or RNA) sequences of different lengths. A general amino acid sequence pattern for zinc fingers is $Cys-X_{2-4}-Cys-X_{12}-His-X_{3-5}-His$ (X is an amino acid) and this can fold about the zinc giving a tetrahedral coordination via the cysteine and histidine donor atoms. Computer searches for related sequences has revealed potential zinc finger domains in several classes of proteins involved in nucleic acid recognition (Fig. 6.5). Two other zinc-containing motifs have been established by three-dimensional NMR studies; both may be regarded as dinuclear. The motif found in the glucocortinoid receptor (GR) has two zinc atoms with an interatomic distance of 13Å, each of which is coordinated tetrahedrally by four cysteines; this motif has been called a zinc twist. That present in the yeast transcription factor GAL4 has the two zinc atoms separated by 3.5Å and there are six cysteine groups, two of which are shared

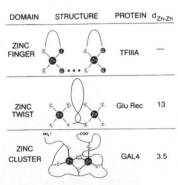

Fig. 6.5. Structural zinc motifs (reproduced with permission from B. L. Vallee).

along a common tetrahedral edge; this has led to this motif being termed a zinc cluster.

6.3 Trinuclear zinc constellations

There is now a growing awareness of the involvement of trinuclear constellations of metal atoms at the active sites of biomolecules as structural information concerning these sites has been made available. The term *constellation* has been introduced by protein crystallographers to describe a cluster of closely spaced metal ions present in a metalloprotein or metalloenzyme. It is an appropriate analogy when one considers the heavy metal atoms standing out against a light atom background and thinks of the sky at night. It is preferred here to 'cluster', as cluster has acquired a particular meaning in chemistry. A cluster is defined as a group of two or more metal atoms in which there are substantial and direct bonds between the metal atoms as in species such as $[Re_3Cl_{12}]^{3-}$ and $Ru_3(CO)_{12}$.

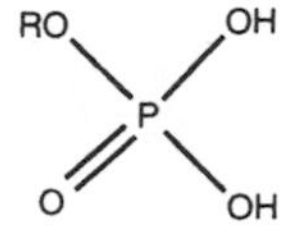

The first zinc enzyme to be discovered in which there are three metal atoms (two zinc and one magnesium, or in the absence of magnesium, a third zinc) at the active site was alkaline phosphatase (AP), which is a dimeric metalloenzyme having phosphomonoesterase activity. The X-ray crystal structure of AP complexed with inorganic phosphate (the E•P complex) was determined by Harold Wyckoff and shows that the three metal atoms are in close proximity (Fig. 6.6). The intermetallic separations are $d(Zn_1–Zn_2) = 3.94\text{Å}$, $d(Zn_2–Mg) = 4.88\text{Å}$, and $d(Zn_1–Mg) = 7.09\text{Å}$ in one subunit and 4.18, 4.66, and 7.08Å respectively in the second subunit.

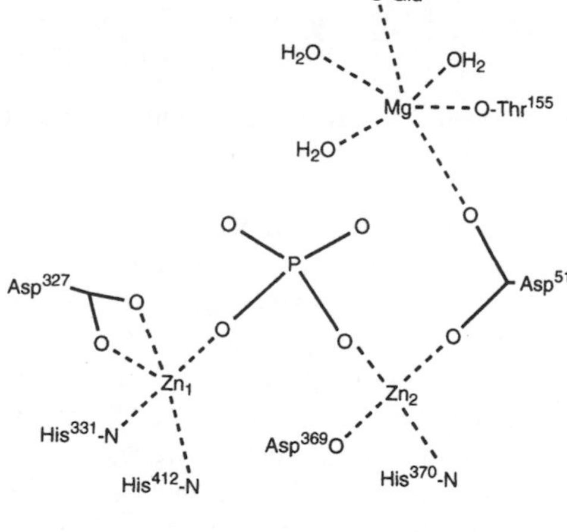

phosphatemonoester

Fig. 6.6. Schematic of the metal constellation in the *Escherichia coli* AP–inorganic phosphate complex.

The first zinc atom (Zn_1) is pentacoordinated and the coordination polyhedron at the metal is best described as pseudotetrahedral with both oxygen atoms of the chelating Asp-327 occupying one apex; the second zinc atom (Zn_2) is tetrahedrally coordinated and the zinc atoms are bridged by the inorganic phosphate. The carboxyl group of Asp-51 forms a bridge between Zn_2 and the magnesium atom, and the phosphate is also associated with the magnesium atom via one of the water molecules coordinated to that atom. The magnesium atom has a slightly distorted octahedral coordination environment.

Interestingly all three types of carboxylate-to-metal coordination are found at the trinuclear site. The unidentate mode of bonding is exhibited, for example, by Asp-369 (to Zn_2), the bidentate mode is found with Asp-327 binding Zn_1 and the bridging mode is found with asp-51 bridging Mg and Zn_2. This phenomenon has also been noted in the structure of the non-haem diiron metalloenzyme ribonucleotide reductase.

Related homotrinuclear zinc constellations have been found in the enzyme phospholipase C (Fig. 6.7) and the P1 nuclease from *Penicillium citrinium*

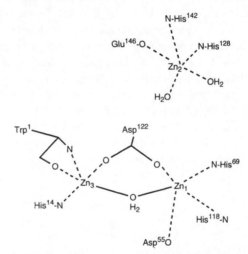

Fig. 6.7. Schematic of the metal constellation in phospholipase C from *Bacillus cereus*—the site in the P1 nuclease from *Penicillium citrinium* is closely similar.

—this nuclease is an endonuclease which hydrolyses single stranded ribo- and deoxyribonucleotides. The active sites are closely similar and the dinuclear centre differs from that in AP as it is doubly bridged by a bidentate aspartate residue and by a molecule of water or the hydroxyl group.

The sites in phospholipase C and P1 nuclease are all typical zinc sites built up from nitrogen and oxygen donors, whereas in AP only two sites are of this form. The third site is an all-oxygen donor site and so this may explain its preference for magnesium. The aspartate group bridging pattern (Fig. 6.8) is also different as it bridges two zinc atoms and so defines a dinuclear zinc site in phospholipase C and P1 nuclease but bridges magnesium and zinc in AP in which the two non-bridged zincs are sufficiently close to provide a functional dinuclear pair bridged by the phosphate.

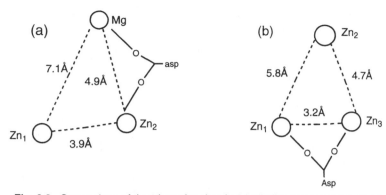

Fig. 6.8. Comparison of the triangular sites in (a) alkaline phosphatase and (b) phospholipase C.

A mechanistic proposal for phosphomonoesterase activity has been presented which suggests that it functions through a phosphoseryl intermediate to produce free inorganic phosphate, or to transfer the phosphoryl group to other alcohols (Fig. 6.9). The left-hand side of the scheme shows the formation of a phosphoseryl intermediate (E–P) and the right-hand side shows the hydrolysis of it to release inorganic phosphate (P_i).

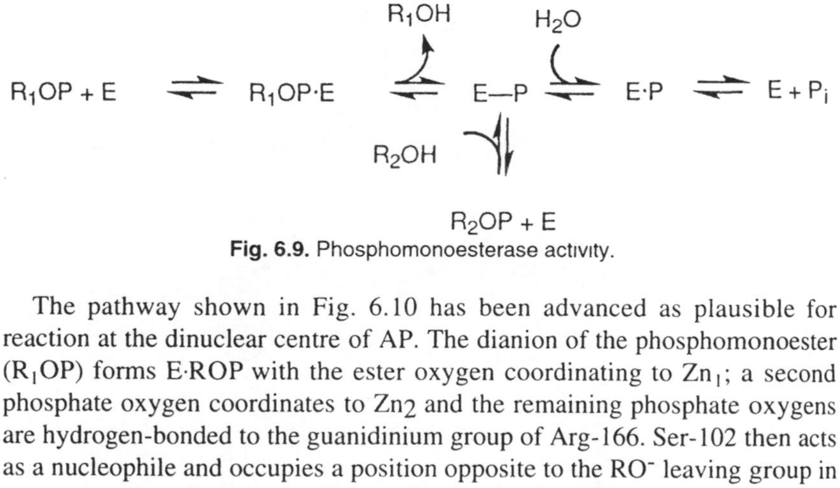

$$R_1OH \qquad H_2O$$

$$R_1OP + E \;\rightleftharpoons\; R_1OP{\cdot}E \;\rightleftharpoons\; E{-}P \;\rightleftharpoons\; E{\cdot}P \;\rightleftharpoons\; E + P_i$$

$$R_2OH$$

$$R_2OP + E$$

Fig. 6.9. Phosphomonoesterase activity.

The pathway shown in Fig. 6.10 has been advanced as plausible for reaction at the dinuclear centre of AP. The dianion of the phosphomonoester (R_1OP) forms E·ROP with the ester oxygen coordinating to Zn_1; a second phosphate oxygen coordinates to Zn_2 and the remaining phosphate oxygens are hydrogen-bonded to the guanidinium group of Arg-166. Ser-102 then acts as a nucleophile and occupies a position opposite to the RO⁻ leaving group in a five-coordinate intermediate. Upon formation of E–P a water molecule can coordinate to Zn_1 in the position vacated by RO⁻; at alkaline pH this water loses a proton to give Zn–OH which acts as a nucleophile for the hydrolysis of the phosphoseryl ester. The E·P complex forms and this is followed by dissociation of the phosphate from this complex.

The two metal atoms therefore activate the oxygens of the water and Ser-102, and also stabilize the appropriate leaving groups. Although the magnesium site does not appear to play any direct role in the catalysis steps it is known that its presence does enhance the catalytic ability of the enzyme. It may therefore be that the magnesium plays a structural role by moderating the necessary environment for catalytic activity through the hydrogen-bonded network associated with the magnesium-coordinated water molecules.

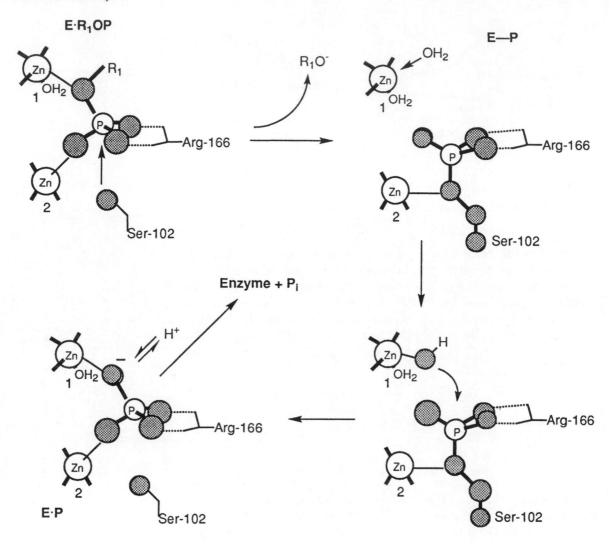

Fig. 6.10. Mechanistic proposal for for AP activity (after Coleman).

Such two-metal-assisted catalysis might be a more general phenomenon and applicable not only in the zinc constellations but also in inorganic pyrophosphatases which hydrolyse inorganic pyrophosphate (PP_i) to inorganic phosphate (P_i) and are known to include at least three manganese atoms in close proximity.

6.4 Vitamin B_{12}—nature's organometallic compound

Although it had been known for since the 1920s that a diet containing raw liver could stimulate the regeneration of red blood cells in anaemic mammals it was not until 1948 that the so-called 'pernicious anaemia factor' of the raw liver was actually isolated. It was crystallized independently by research teams in the USA (Merck Laboratories) and in the UK (Glaxo Laboratories).The dark red crystals were given the name vitamin B_{12} and subsequently shown to be cyanocobalamin, a diamagnetic, six-coordinate cobalt(III) complex in which the cyano group is a result of the isolation procedure.

In 1958 the coenzyme form of vitamin B_{12} (subsequently referred to as B_{12}) was isolated. Its crystal structure revealed the presence of two structural components previously unobserved in nature, the corrin ring and a cobalt–carbon bond making it the first naturally occurring organometallic compound (Fig. 6.11) to be identified.

Dorothy Crowfoot Hodgkin was awarded the Nobel Prize for Chemistry in 1964 for her determination by X-ray techniques of the structures of important biochemical molecules including B_{12} and the coenzyme.

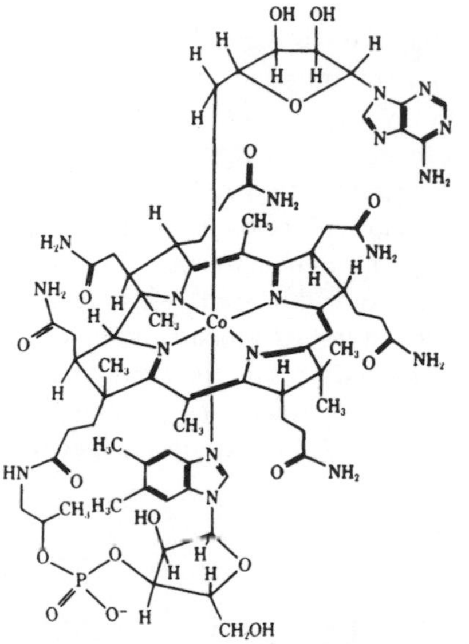

Fig 6.11. Vitamin B_{12} coenzyme.

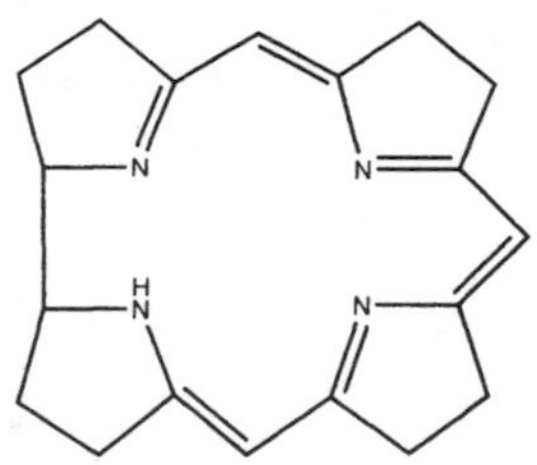

Corrin framework

The corrin ring is similar to a porphyrin ring but has one methine (-CH=) group less. It provides a square planar set of nitrogen atoms and access for axial groups above and below its plane. In naming compounds derived from B_{12} the presence of the associated 5,6-dimethylbenzimidazole group as one axial group leads to the name cobalamin, and the occupant of the second axial site determines the prefix used with it hence B_{12} is cyanocobalamin and the coenzyme can be designated 5'-deoxyadenosylcobalamin. The aquo complex is B_{12a} and can be readily reduced to B_{12r} containing Co(II) and B_{12s} containing Co(I). These reductions may be carried out by NADH or FAD (flavin adenine dinucleotide) respectively, or by chemical methods such as catalytic hydrogenation or reduction by chromium(II) acetate at pH 5 for the former or by sodium borohydride or reduction by chromium(II) acetate at pH 9.5 for the latter.

The Co(I) derivative is a powerful nucleophile and undergoes alkylation by oxidative-addition:

$$[B_{12}Co(I)] + CH_3I \rightarrow [B_{12}Co(III)CH_3]^+ + I^-$$

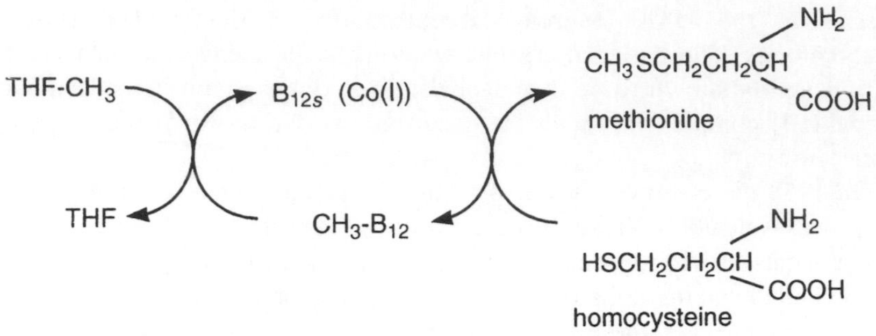

Fig. 6.12. The conversion of homocysteine to methionine.

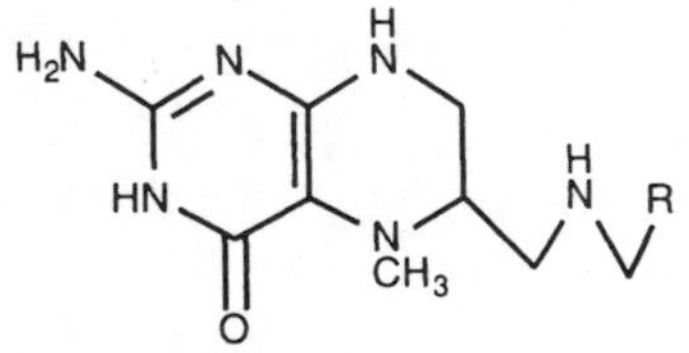

N^5-methy-tetrahydrofalate

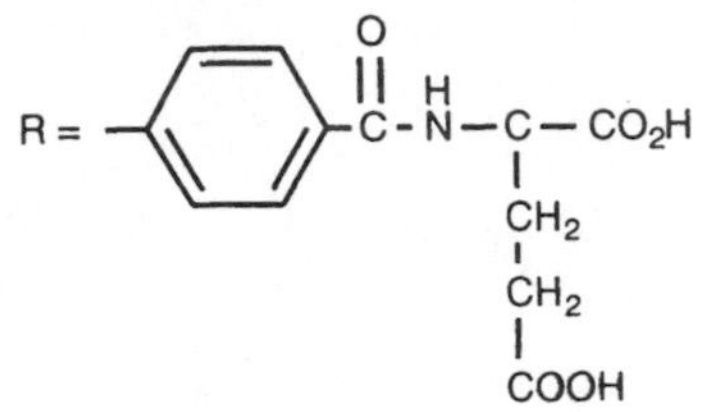

An important function of B_{12} is the acceptance of a methyl group from N^5-methyltetrahydrofolate to produce methylcobalamin which can then participate in biomethylation reactions such as the terminal step of methionine biosynthesis in which homocysteine is converted into methionine (Fig. 6.12). Methyl transfer from methylcobalamin requires cleavage of the cobalt–carbon bond. Under different conditions this then gives a carbanion (CH$_3^-$), a radical (CH$_3$·), or a carbonium ion (CH$_3^+$).

A second class of enzymic reactions involving B_{12} have 5'-deoxyadenosylcobalamin as the prosthetic group. The reaction of B_{12s} with ATP (adenosine triphosphate) results in the formation of a cobalt–carbon bond between adenosine and cobalt, and formation of the coenzyme. It is very effective in inducing 1,2-shifts of the general type which are of importance in

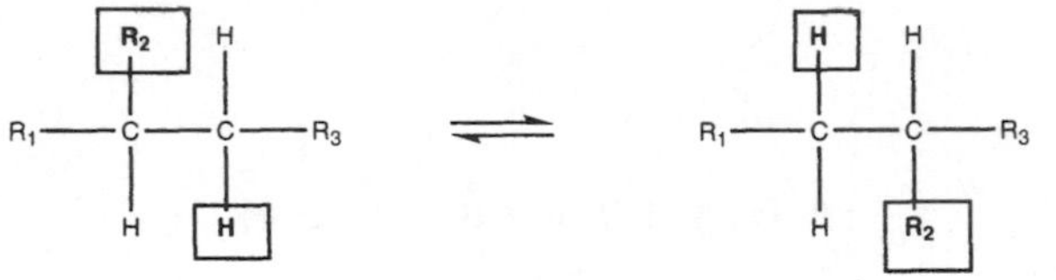

metabolism (the migrating groups are depicted in boxes). A representative group of these reactions are shown in Table 6.2, and illustrate C–C, N–C and O–C bond cleavage. The shift is followed by an internal condensation

Table 6.2. Representative 1,2-shift reactions

	R$_1$	R$_2$	R$_3$
Diol dehydratase	CH$_3$	OH	OH
Ethanolamine deaminase	H	NH$_2$	OH
Glutamate mutase	H	CH(NH$_2$)COOH	COOH
Methylmalonyl CoA mutase	H	CO-CoA	COOH
Glycerol dehydratase	CH$_2$OH	OH	OH

reaction to give the final product; for example, propane-1,2-diol would yield propanal, and ethanolamine would give ethanal.

The mechanism of the reaction is postulated as occurring via a radical pathway and one formulation is depicted in Fig. 6.13. Homolytic cleavage of the Co–deoxyadenosine bond yields B_{12r} and the deoxyadenosyl radical which abstracts hydrogen from the substrate to give 5'-deoxyadenosine and the substrate radical. The relative importance of the factors enhancing homolytic cleavage is still not determined. Rearrangement of the substrate radical takes place, possibly with enzymic intervention, and then abstraction of hydrogen from 5'-deoxyadenosine places hydrogen in its new position in the product. The deoxyadenosyl radical formed reacts with B_{12r} to regenerate the coenzyme.

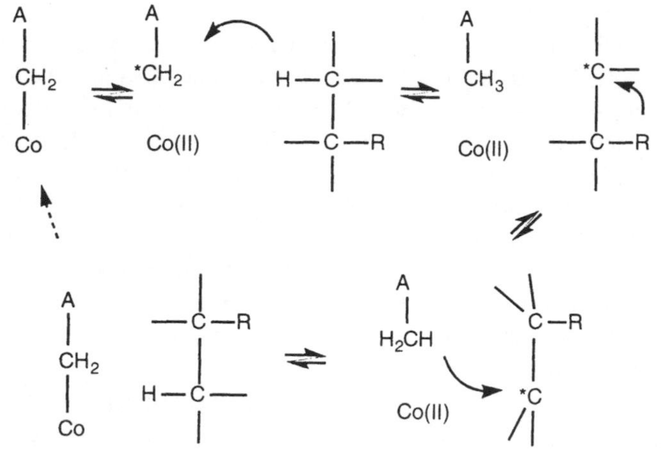

Fig. 6.13. Generalized 1,2-shift reaction (the free radicals are asterisked).

The biomethylation of mercury

The toxic effects of mercury have been long known. The Mad Hatter in Lewis Carroll's *Alice in Wonderland* illustrates the nervous disorder produced by exposure to mercury salts in the curing of the pelts from which hats were made. Organomercurials such as methylmercury are more dangerous than mercury metal or inorganic mercury salts as they have longer half-lives to accumulation in the body—70 days for methylmercury as compared with six days for inorganic mercury—and so can reach 10 times the level of inorganic mercury after nine months. This means that even the most protected areas of the body such as the brain and, during pregnancy, the foetus become vulnerable to attack.

There have been major instances of mercury poisoning. In Minimata, Japan, in 1953, 52 people died after eating sea-food contaminated with mercury, and in 1971–72 in Iraq, hundreds of people died and thousands more were seriously affected by eating bread made from a relief shipment of grain that had been dressed with an organomercurial fungicide. In Sweden, effects have been noted as a consequence of mercury-containing effluent from paper mills.

In 1964 Jack Halpern reported methyl transfer to mercury(II) species from pentacyanomethyl cobaltate and suggested that there was a similarity between this reaction and B_{12} activity. In 1968 John Wood demonstrated enzymic transfer to mercury(II) from methylcobalamin and so the link had been made. Naturally occurring anaerobic bacteria in the sediments of the sea, rivers, and lakes can alkylate mercury which is then taken up by plankton and so enters into the food chain with progressive concentration.

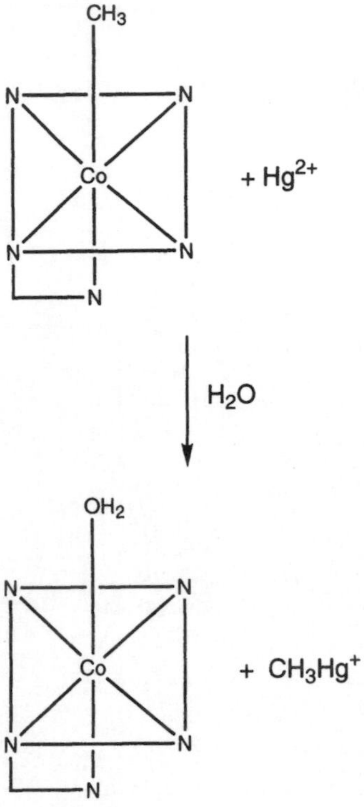

The overall mechanism of the reaction is shown in Fig. 6.14 and, as a carbanion is most likely to react with a positively charged metal, CH_3^- is the electrophilic transfer agent. It is possible that the reaction proceeds via a fast stage in which the apical base reacts with mercury and is displaced from the cobalt, followed by alkyl transfer and release of the organomercurial accompanied by generation of B_{12a}. It is now known that other metals such as tin, lead, thallium, and palladium can also be methylated by methylcobalamin.

Once mercury is in the body, methyl mercury chloride is synthesized in the gut; the relatively non-polar methyl mercury is lipid-soluble and so is easily transported across biomembranes into the blood stream. It then reaches the brain and central nervous system where it has a devastating effect. It has been speculated that this occurs because of the solubility of methyl mercury in phospholipids which enables lysis of certain cell membranes and hence cell breakdown. The proposal is enhanced by the observation from NMR studies on the interaction of methylmercury chloride with plasmalogens, phospholipids which contain an α,β-unsaturated ether linkage, which are important in membrane structure of cells of the central nervous system; exposure to the mercurial caused hydration and hydrolytic cleavage of the plasmalogen (Figure 6.15).

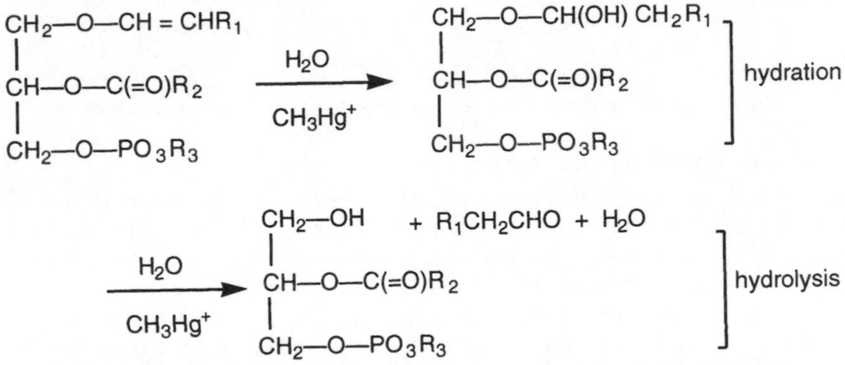

Fig. 6.14. The alkylation of Hg^{2+}.

$$CH_2-O-CH=CHR_1 \qquad \xrightarrow[CH_3Hg^+]{H_2O} \qquad CH_2-O-CH(OH)\,CH_2R_1$$
$$CH-O-C(=O)R_2 \qquad \qquad CH-O-C(=O)R_2 \qquad \text{hydration}$$
$$CH_2-O-PO_3R_3 \qquad \qquad CH_2-O-PO_3R_3$$

$$\xrightarrow[CH_3Hg^+]{H_2O} \qquad CH_2-OH \quad + \; R_1CH_2CHO \; + \; H_2O$$
$$CH-O-C(=O)R_2 \qquad \text{hydrolysis}$$
$$CH_2-O-PO_3R_3$$

Fig. 6.15. Proposed mechanism for the methylmercury-catalysed lysis of plasmalogen (L-α-phosphatidyl ethanolamine). [R_1 = a mixture of stearate and palmitate, R_2 = linolenate; $R_3 = NH_2CH_2CH_2O^-\cdot$]

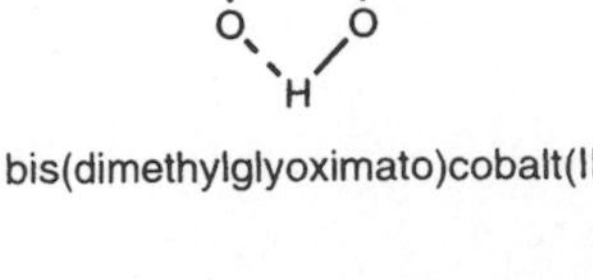

bis(dimethylglyoximato)cobalt(II)

Although no stable cobalt alkyls were known until the elucidation of the structure of the B_{12} coenzyme, many are now available as certain simple coordination complexes of cobalt such as bis(dimethylglyoximato)cobalt(II), and related complexes, and the tetradentate Schiff base complexes used to study dioxygen activity readily form alkyl derivatives. These have been used to mimic the reactions of B_{12} and to explore the range of metal alkylation available.

6.5 Nitrogenases

It is amusing to think that while man labours to reduce dinitrogen to ammonia using an energetically expensive process requiring high temperature, pressure, and a promoted iron catalyst, nature, can fix two to three times as much dinitrogen each year. Nature uses a wide range of bacteria in nitrogen fixation, the best known of which is *Rhizobium,* found in nodules in the roots of legumes such as peas, beans, clover, and soya.

Nitrogen fixation is one of nature's fundamental processes and is catalysed by nitrogenases. The Fe–Mo nitrogenases are those that have been most studied and it has been traditionally believed, although the evidence was circumstantial, that reduction of dinitrogen took place at the molybdenum site. It is now known that there are Fe–V and Fe–Fe nitrogenases. The discovery of the former showed that the purported role of the molybdenum might also be carried out by other metals and the isolation of the latter clearly indicated that it was not actually necessary to have different metals present in the enzyme in order to effect fixation.

Fe–Mo nitrogenases

The Fe–Mo nitrogenases consist of two proteins: an Fe protein, of molecular weight *c.* 60,000, which is a dimer of two identical subunits bridged by an $[Fe_4S_4]$ cluster; and an Fe–Mo protein, having molecular weight *c.* 220,000, which is a tetramer containing two Mo atoms, 30–32 iron atoms and 30–32 inorganic sulphur atoms (Fig. 6.16).

The direct combination of N_2 and H_2 using a promoted iron catalyst is called the Haber process. It requires a high temperature (~450°C) to overcome the kinetic inertness of N_2 and a high pressure (~270 atm) to overcome the thermodynamic effect of an unfavourable equilibrium constant at 450°C. A promoted iron catalyst consists of iron with small amounts of oxides such as MgO, Al_2O_3 or SiO_2. Two Nobel Prizes are associated with this process. In 1918 Fritz Haber received the award for Chemistry for his development of the process and in 1931 Carl Bosch was awarded the Nobel Prize for Chemistry for overcoming the chemical engineering problems associated with the use of large-scale high-pressure techniques.

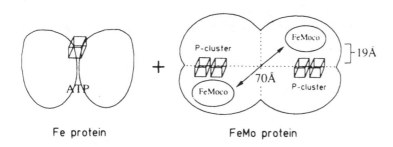

Fe protein   FeMo protein

Fig. 6.16. A schematic representation of Fe–Mo nitrogenase based on the X-ray structure analysis of the Fe and Fe–Mo proteins. [Cube = Fe_4S_4 and FeMoco = FeMocofactor.] (Reproduced with permission from Angewandte Chemie.)

The reaction catalysed is:

$$N_2 + 8H^+ + 8e^- + 16MgATP \rightarrow 2NH_3 + H_2 + 16MgADP + 16PO_4^{3-}$$

and the pathway proposed for the reaction was that the electrons required for the reaction are transferred to nitrogenase by reduced forms of ferredoxins and flavodoxins (Fig. 6.17); the source of these electrons is the oxidation of pyruvate (2-oxo-propionate). The electrons are first transferred to the Fe protein, the reduced form of which forms a complex with monomagnesium

When N_2-fixing bacteria are grown on iron-deficient media, flavodoxin is produced. A flavodoxin is a low molecular weight flavoprotein containing neither iron nor labile sulphar but having ribflavin mononucleotide present as the prosthetic group. Flavodoxins and ferrodoxins appear to be interchangeable with regard to their activity in N_2-fixation.

ATP → ADP + P$_i$

ΔG^o = —30.5 kJmol^{-1}

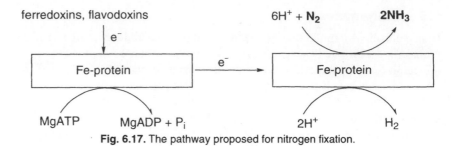

Fig. 6.17. The pathway proposed for nitrogen fixation.

ATP and the FeMo protein. The reducing electron is then transferred to the FeMo protein thence to the dinitrogen and in a series of steps accompanied by proton transfer from water to the dinitrogen produces ammonia, monomagnesium ADP, and inorganic phosphate; the proteins are regenerated in their original states. Although it has been suggested that the energy released by the hydrolysis of ATP drives the reaction, the exact role of the magnesium ATP is not yet fully understood.

The architecture of the FeMo protein has posed a major chemical challenge. Extrusion experiments revealed that the FeS centres contain [Fe$_4$S$_4$] clusters but the spectroscopic properties of these clusters were so different from those of known [Fe$_4$S$_4$] clusters that the clusters were termed 'unusual P clusters'. Numerous speculative small molecule models derived from iron–sulphur clusters in which an atom of iron had been replaced by a molybdenum, were constructed to mimic the anticipated structural features of the Fe–Mo site but none of them have shown any reactivity towards dinitrogen (Fig. 6.18). It is interesting at this point to note a recent comment that rich veins of metal-dinitrogen and metal-chalcogenide chemistry have been laid down by synthetic chemists in trying to obtain the answer to the structure of the nitrogenase cluster, and to remember that for many generations dinitrogen was regarded as chemically inert.

The solution of the X-ray crystal structure of the Fe–Mo protein from *A. vinelandii,* by Douglas Rees and his group, shows that the two Fe–Mo cofactors are located some 70 Å apart with the 'P clusters' 19Å from them. The structures of the Fe–Mo cofactor and the 'P cluster' pair are shown in Fig. 6.19. The Fe–Mo cofactor consists of cuboidal [Fe$_4$S$_3$] and [Fe$_3$MoS$_3$] fragments which are linked by two sulphide bridges and a third bridge which might derive from an O or N donor. The cluster is bound to the protein via a cysteine at Fe(1) and a histidine at the Mo. The coordination sphere of the six-coordinate molybdenum, which is probably in oxidation state +IV, is completed by the homocitrate anion. The cluster differs from all of the small molecule models that have been proposed, particularly in the threefold linkage of the two cuboidal fragments. The 'P cluster' pair contains [Fe$_4$S$_4$] units linked by two bridging cysteine thiolate groups, together with a disulphide bridge formed between sulphur atoms from each [Fe$_4$S$_4$] cubane cluster.

The structure proposed has been said to account for the observed physico-chemical properties of the Fe–Mo protein but there remains the intriguing question of how does it work and at which site does the dinitrogen bind? The

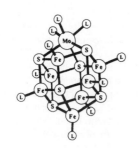

M=Mo. V or Fe

Fig. 6.18 . Representative models for the Fe–M–S site in nitrogenase (reproduced with permission from the American Chemical Society).

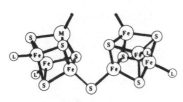

CH$_2$COOH

C(OH)COOH

CH$_2$

CH$_2$COOH

homocitric acid
[2-hydroxy-1,2,4-butane tricarboxylic acid]

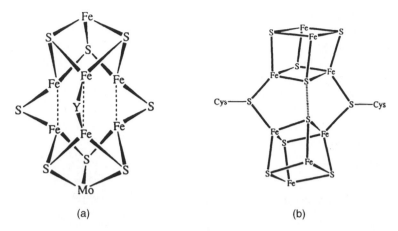

Fig. 6.19. The structure of (a) the Fe–Mo cofactor and (b) the 'unusual P cluster' in the Fe–Mo protein of nitrogenase (reproduced with permission from *Science*).

long-held view that the molybdenum is the active centre for dinitrogen bonding has, as commented earlier, lost favour but although the molybdenum atom in this resting state of the enzyme is coordinatively saturated and so not predisposed towards binding the dinitrogen it is not inconceivable that on reduction the metal, now 'softer', would lose the hard homocitrate ligand and so produce free sites for dinitrogen binding. If not the Mo then at which Fe centre or centres is there interaction? One speculative model has been proposed for the functioning of nitrogenase in which dinitrogen binds to a pair of privileged iron atoms that have open coordination sites, in a 'side-on' manner reminiscent of the $\eta^2:\eta^2$ interaction observed for dioxygen in oxyhaemocyanin (**6.3**). It is also possible to consider that a simple bridge could occur at the same site (**6.4**). There are many questions still to answer and many challenges to meet. The principal challenge for the synthetic chemist is to produce a small molecule model, based on the above structural proposition, that has the potential to bind dinitrogen.

6.3

6.4

6.6 Oxotransfer molybdoenzymes

Molybdenum is the only element from the second- and third-row transition metals that has been found to be essential to life. As well as functioning in nitrogenase it has been

Table 6.3 . Some oxomolybdenum enzymes

Enzymes	Source	Components
Formate dehydrogenase	Fungi, bacteria, plants	Mo, Se, Fe_nS_n
Carbon monoxide oxidase	Bacteria	2Mo, $4Fe_2S_2$, 2FAD, 2Se
Nitrate reductase	Fungi, algae, bacteria, plants	2Mo, $2Cyt_b$, 2FAD
Xanthine oxidase	Cow's milk, mammalian livers	2Mo, $4Fe_2S_2$, 2FAD
Aldehyde oxidase	Pig's liver	2Mo, $4Fe_2S_2$, 2FAD
Sulphite oxidase	Mammalian livers, bacteria	2Mo, $2Cyt_b$

found to be active in oxotransfer enzymes which catalyse the transfer of oxygen atoms to, or from, a substrate and are involved in carbon, nitrogen, and sulphur metabolism (Table 6.3).

Although a crystal structure of an oxomolybdenum enzyme is not yet available a systematic study of such sites using EXAFS spectroscopy and comparisons with model coordination complexes has revealed that there is a structural diversity available.

A common structural feature emerging from the EXAFS-determined sites is the presence of an Mo=O bond; the metal sites in the oxidized and reduced forms of xanthine oxidase, sulphite oxidase, and *Chlorella* nitrate reductase are shown below.

$$
\begin{array}{ll}
\text{Sulphite} & (RS)_{2,3}\!-\!\overset{\displaystyle O}{\overset{\|}{Mo(VI)}}\!=\!O \qquad (RS)_3\!-\!\overset{\displaystyle O}{\overset{\|}{Mo(IV)}} \\
\text{oxidase} & \qquad\qquad \text{oxidized} \qquad\qquad\quad \text{reduced}
\end{array}
$$

$$
\begin{array}{ll}
\text{Xanthine} & (RS)_2\!-\!\overset{\displaystyle O}{\overset{\|}{Mo(VI)}}\!=\!S \qquad (RS)_2\!-\!\overset{\displaystyle O}{\overset{\|}{Mo(IV)}}\!-\!SH \\
\text{oxidase} & \qquad\qquad \text{oxidized} \qquad\qquad\quad \text{reduced}
\end{array}
$$

$$
\begin{array}{ll}
\text{Nitrate} & (RS)_2\!-\!\overset{\displaystyle O}{\overset{\|}{Mo(VI)}}\!=\!O \qquad (RS)_{3,4}\!-\!\overset{\displaystyle O}{\overset{\|}{Mo(IV)}} \\
\text{reductase} & \qquad\qquad \text{oxidized} \qquad\qquad\quad \text{reduced}
\end{array}
$$

The oxotransfer molybdoenzymes also have present dissociable cofactors which are obligatory for enzyme action. These cofactors are small molecules containing molybdenum which can be released from one protein and then introduced into a second apo-protein structure to form an active enzyme. For example there is a nitrate reductase mutant, Nit-1, from the fungus *Neurospora crassa* which contains all of the components shown in the above table but has no Mo and no Mo-cofactor. If the Mo-cofactor from another oxotransfer enzyme is introduced into Nit-1 then this produces the intact and active nitrate reductase. The cofactor has been shown to include a novel pterin, the properties of which depends on the nature of the side chain. The proposed structure for the molybdopterin (**6.5**) suggests that the molybdenum is bound to the chelating dithiolene function.

Many attempts have been made to produce model compounds for the oxo-molybdenum sites, and also to reproduce key features of the molybdenum cofactor. Many of these have involved complexes derived from the *cis*-MoO_2 core and thiolato ligands but although these studies have frequently revealed novel aspects of molybdenum chemistry, the models have not yet duplicated nature. Some general features of oxomolybdenum chemistry (Fig. 6.20) that

6.5

have emerged from the model studies include oxygen atom addition or removal especially at terminal sites, conproportionation reactions between Mo(IV) and Mo(VI) species to give monooxygen-bridged dinuclear Mo(V) species and the formation of thermodynamically stable dimers. As there is no evidence for dimers at the active sites then one synthetic challenge is to prevent their formation in small molecule models and so large multidentate ligands have been devised to do this. The tetradentate ligand shown (**6.6**) forms a mononuclear Mo(VI) complex whose potential dimerization on reduction is sterically hindered and the complex formed from the tridentate ligand (**6.7**) remains monomeric during oxygen-transfer reactions from triphenylphosphine (PPh₃) to triphenylphosphine oxide (PPh₃O) and from dimethyl sulphide to dimethylsulphoxide.

The oxotransfer enzymes catalyse either oxidations or reductions and these may be written either as formal oxygen atom transfer reactions ($NO_3^- \rightarrow NO_2^- + [O]$; $SO_3^{2-} + [O] \rightarrow SO_4^{2-}$) or as redox half-reactions where protons, electrons, and water molecules all participate in the reaction ($NO_3^- + 2H^+ + 2e^- \rightarrow NO_2^- + H_2O$; $SO_3^{2-} + H_2O \rightarrow SO_4^{2-} + 2H^+ + 2e^-$). These processes

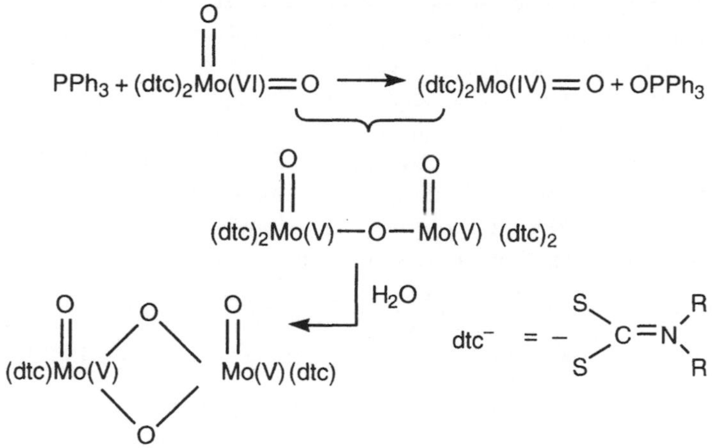

Conproportionation is the reverse of disproportionation; two substances containing the same element having different oxidation numbers form a product in which the element has an intermediate oxidation number.

Fig. 6.20. Oxotransfer by molybdenum dithiocarbamate complexes

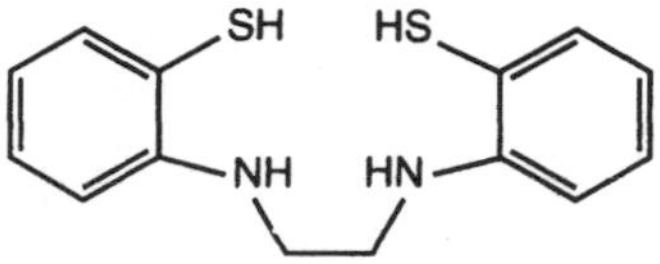

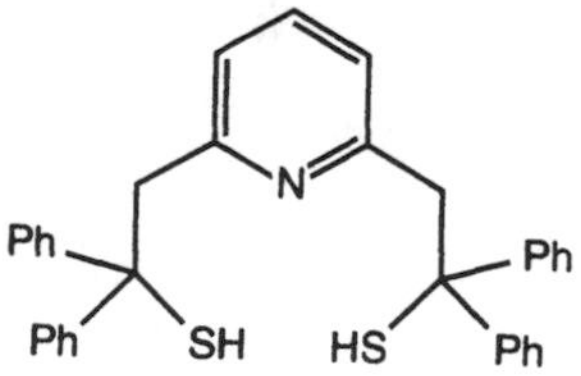

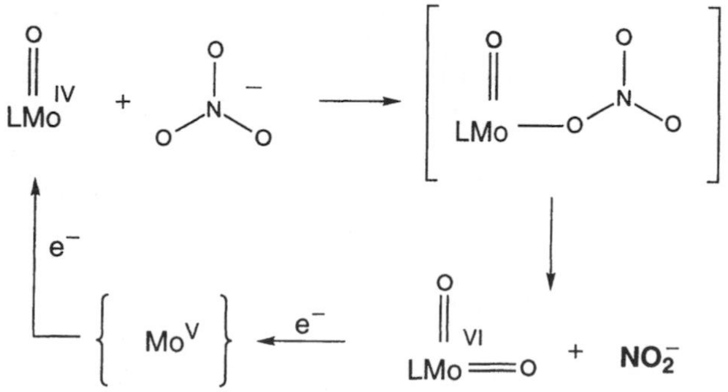

Fig. 6.21. Nitrate reduction by oxygen-atom transfer.

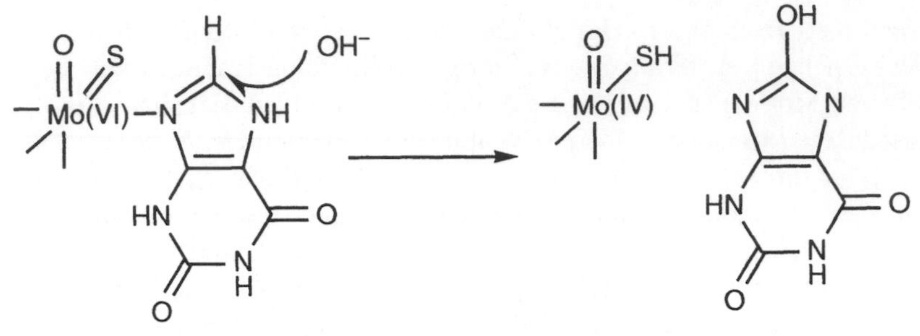

Fig. 6.22. Xanthine oxidation by proton-coupled electron transfer.

are illustrated here by reference to oxygen atom transfer at the molybdenum site in nitrate reductase (Fig. 6.21) and a proton-coupled electron transfer at the molybdenum site in xanthine oxidase (Fig. 6.22).

6.7 Nickel enzymes

In recent years a number of plant species have been discovered which can accumulate high concentrations of nickel. It is therefore perhaps surprising to find that although nickel is a relatively abundant metal (0.008% of the earth's crust) which is available to organisms through leaching, no protein containing functionally significant nickel was known until 1975. It is also quite ironic that the first enzyme to be crystallized, by James B. Sumner in 1926, was urease from jack beans as this urease was found, some fifty years later, to contain nickel—so becoming the first known nickel enzyme. Nickel has subsequently been found in several different systems.

The discovery that enzymes could be crystallized led to James B. Sumner being awarded a share in the 1946 Nobel Prize for Chemistry.

Urease

Urease catalyses the hydrolysis of urea to ammonia and carbonic acid,

$$(NH_2)_2C{=}O + 2H_2O \rightarrow 2NH_3 + H_2CO_3$$

The discovery, by Burt Zerner and his research group, of the presence of nickel came from a critical analysis of the UV-visible spectrum which showed a distinct long wavelength absorption characteristic of octahedral nickel(II) in an oxygen- and nitrogen-donor environment. EXAFS studies are consistent with this assignment and the enzyme has been found to contain two nickel ions which have been postulated to act cooperatively. Each has a different role in the mechanism of reaction with one serving to polarize the urea and the second enhancing the nucleophilicity of water so that the hydroxide formed can attack the polarized carbonyl of the urea (Fig. 6.23). The reaction is analogous to that of peptide hydrolysis using the zinc metalloenzyme carboxypeptidase.

The crystal structure of urease has revealed that the nickel atoms are 3.5Å apart and bridged by a lysine-derived carbamate bridge. One nickel atom (Ni_A) is 5-coordinate, being coordinated by two histidines, one aspartate, the bridging carbamate and a water molecule, and the second nickel atom (Ni_B) is three-coordinate being linked to the bridging carbamate and two histidines. Ni_B is coordinatively unsaturated and so could host the incoming urea and polarize the carbonyl group for nucleophilic attack.

Nickel-containing hydrogenases and CO dehydrogenases

Several bacterial species have present hydrogenases which catalyse the reaction below

$$H_2 \rightarrow 2H^+ + 2e^-$$

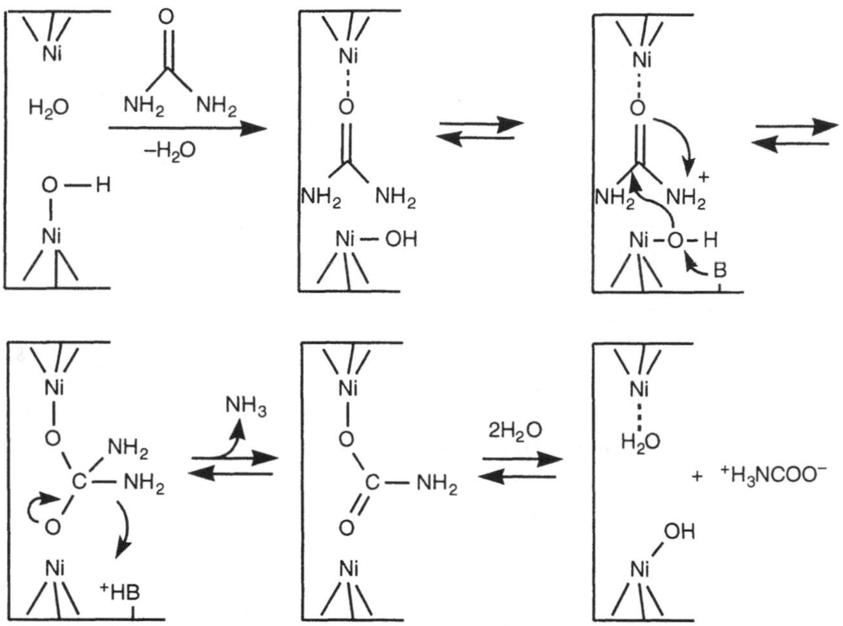

Fig. 6.23. A mechanistic proposal for the action of urease (after Blakely).

All known hydrogenases contain Fe–S clusters and many, but not all, have now been found to contain nickel as well. EPR studies on *Desulphovibrio vulgaris* and methanogenic bacteria have given signals which are attributed to the unusual nickel(III) oxidation state (low spin, d^7); furthermore the ease of obtaining nickel(III) suggests that the coordination geometry might be octahedrally based. The type of ligand coordination which leads to this state is not yet fully established but X-ray absorption studies on the hydrogenase isolated from the bacterium *Thiocapsa roseopersoicina* suggest that the coordination sphere of the metal comprizes $3(\pm 1)$ nitrogen or oxygen donors and $2(\pm 1)$ sulphur donors. If the nickel is the catalytic site for the oxidation of dihydrogen then the Fe–S clusters will serve as one-electron oxidants for the nickel.

Nickel-containing CO dehydrogenases or carbon monoxide oxido-reductases act to interconvert carbon monoxide and carbon dioxide, as shown below.

$$CO + H_2O \rightarrow CO_2 + 2H^+ + 2e^-$$

These metalloenzymes have been found to contain Fe–S clusters and nickel(III). The Fe–S clusters are derived from 4Fe–4S* clusters and are oxidized by carbon dioxide and reduced by carbon monoxide. It is quite possible that a heteronuclear cluster containing three iron atoms and one nickel atom may be present with the cluster being constructed from a voided 3Fe–4S* cube to which a nickel atom has added.

Methanogens are anaerobic bacteria which convert carbon dioxide to methane using reducing agents such as dihydrogen gas.

Coenzyme F$_{430}$

Coenzyme F$_{430}$ is the cofactor of methylcoenzyme M reductase, which is involved in a complex series of reactions in bacteria leading to the generation of methane gas. It has been identified, by Albert Eschenmoser and his co-workers, as a square planar nickel(II) complex of a tetrahydroderivative of a porphyrinoid related to corrin and hence described as a corphin (Fig. 6.24). It is required to catalyse the reduction of the methyl group of methylcoenzyme M , 2-(methylthio)ethanesulphonate), to methane.

$$CH_3SCH_2CH_2SO_3^- + H_2 \rightarrow CH_4 + HSCH_2CH_2SO_3^-$$

The exact role of the metal is not yet defined but the accessibility of Ni(I) as shown by EPR experiments suggests that it might be involved in methyl-group transfer, electron-transfer reactions, or both.

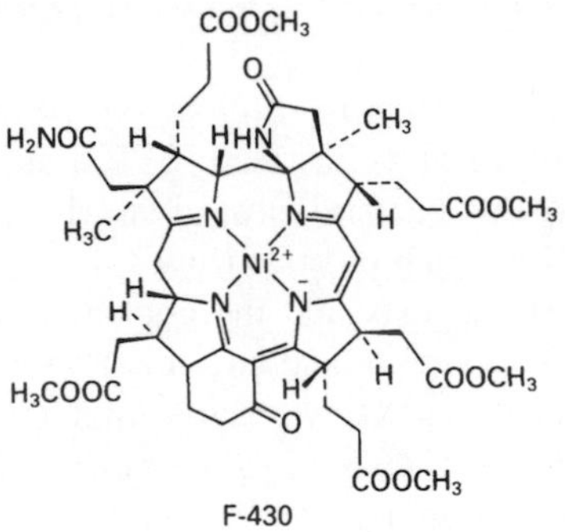

Fig. 6.24. Coenzyme F$_{430}$.

7 Therapeutic uses of coordination compounds

Coordination compounds have found application in medicine in the treatment, management, and diagnosis of disease. In this chapter attention is drawn to two ways in which this occurs—the application of therapeutic chelating agents and the therapeutic use of preformed coordination compounds.

7.1 The application of therapeutic chelating agents

Therapeutic chelating agents have been used to remove excess metals from the body. These might have arisen from metabolic disorders such as Wilson's disease, a hereditary genetic disorder in which the body is unable to metabolize copper in the normal way and copper accumulations occur in the liver and then in the central nervous system, and β-thalassaemia which leads to iron overload and has been discussed in Section 2.2; or by the inadvertent uptake of toxic metals such as lead(II), cadmium, and mercury(II).

The development of chelation therapy is discussed here in terms of the principles used by Sir Rudolph Peters to develope an antidote for the poison gas Lewisite which is an organoarsenic compound that attacks the lungs and the skin. The procedure used followed the following steps, which can be generally used to find a useful chelating agent.

ClCH=CHAsCl$_2$
Lewisite

- Determination of the site of action of the poison (in this case it was found that the arsenic inactivated the –SH groups of enzymes which metabolize pyruvate).
- Determination of the donor group arrangement responsible for binding the metal—the application of the principle of hard and soft acids and bases is useful here.
- Synthesis of chelating agents having the same donor group arrangements.
- Experiments to show that the chelate compound can reverse the enzyme inhibition caused by the metal.
- Critical evaluation of the compound to assess its ability to perform efficiently *in vivo*.

The ideal ligand would be specific for the toxic metal and be non-toxic itself (that is it must possess an LD$_{50}$ ideally greater or equal to 400mg/kg: Table 7.1) and as its complexes. It should form highly stable complexes with the metal ion to be removed that are water-soluble and so readily excretable, and it should not be metabolized. The ligand selected for testing by Peters was 2,3-dimercapto-1-propanol, which subsequently became known as British Anti-Lewisite (BAL) because the chelation of arsenic by the –SH

Table 7.1. Some LD_{50} values (LD_{50} is the statistical estimate of the dosage level of the particular compound which causes a mortality level of 50% in an animal population large enough to treat statistically)

Compound	LD_{50} mg/kg	Species
$Na_2CaEDTA$	3800	Rat
Na_4EDTA	380	Mouse
D-penicillamine	334	Mouse
N-acetyl-D-penicillamine	1000	Rat
Unithiol	2000	Mouse
BAL	105	Rat
Sodium 2,3-dimercaptosuccinate	5000	Mouse

groups with the formation of stable five-membered rings should help remove it from the body (Fig. 7.1). BAL proved effective and was also used for toxic metals such as mercury(II), cadmium(II), and lead(II) which also have an affinity for –SH groups.

As BAL has a degree of toxicity and is difficult to administer a more water-soluble agent was sought. 2,3-Dimercaptosuccinic acid has been shown to be an effective antidote for the same range of metals as BAL, as has sodium 2,3-dimercapto-1-propanesulfonate (unithiol). A further advantage of the latter complexes is that, unlike BAL, they do not facilitate

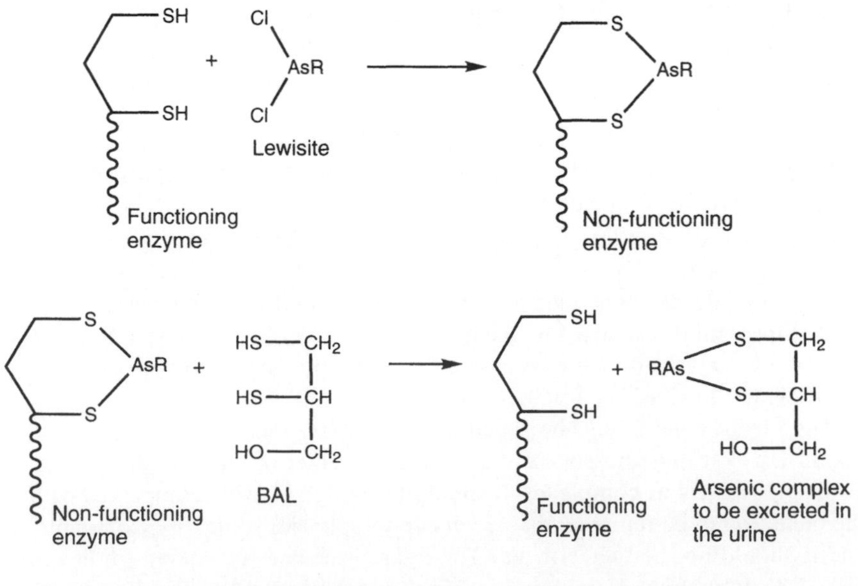

Fig. 7.1. Schematic representation of chelation therapy—here arsenic is used as the example.

the entry of mercury (II) into the brain but concentrate it in the liver and kidneys, making it available for excretion. Cell membranes are constructed of protein and phospholipid, so factors which enhance the lipid solubility of a metal will increase the opportunity for it to cross membranes and enter cells. BAL forms a neutral, lipid-soluble complex with mercury(II) which can readily cross the phospholipid barrier. The mercury(II) complex with unithiol is charged and so remains water-soluble; this inhibits its passage across the membrane and allows for ready excretion.

Polyaminocarboxylic acids have been investigated as therapeutic agents because of their broad chelating ability. Ethylenediaminetetraacetic acid (H_4EDTA) is used here as a representative example. In 1952 the calcium complex of EDTA was shown to protect chicks against lead(II) poisoning.

HS—CH$_2$
HS—CH
NaO$_3$S—CH$_2$

unithiol

HOOCCH$_2$ / N—CH$_2$CH$_2$—N \ CH$_2$COOH
HOOCCH$_2$ CH$_2$COOH

ethylenediaminetetraacetic acid

Later in the same year it was used to treat lead(II) poisoning in humans; the calcium complex was used because H_4EDTA itself can form stable complexes with essential as well as toxic metals and so can interact with and deplete serum calcium. The simple sodium salt is toxic and so the mixed complex Na$_2$CaEDTA is now used as it prevents rapid calcium depletion. In the body, the lead(II) replaces the calcium in the chelating agent and the resulting complex is excreted via the urine. EDTA complexes can be used to remove mercury(II), iron, and other metals from the body but care must be taken that it does not cause depletion of essential elements such as zinc.

D-Penicillamine is a degradation product of penicillin that can utilize nitrogen, oxygen, or sulphur atoms as donors. It is an effective chelating agent for lead(II), mercury(II), and copper(II) and for many other metals. N-acetyl-D-penicillamine is less toxic than D-penicillamine and both reagents can be administered orally. D-Penicillamine has been used in the treatment of rheumatism and arthritis and is the preferred treatment for Wilson's disease.

penicillamine

N-acetylpenicillamine

Both penicillamines have been used against mercury with the N-acetyl derivative being very effective at removing methylmercury(II) from brain tissue. Caution must be exercised with the use of both BAL and D-penicillamine with cadmium(II) as the complexes formed are more toxic than the metal.

The natural approach to chelation therapy has already been demonstrated with the use of desferrioxamine B and related siderophores and analogues in the relief of iron overload (Section 2.2). These systems have some problems, however, as hydroxamates are susceptible to the acid environment of the stomach and the catecholates form charged complexes which can trap the iron intracellularly and so prevent excretion. A new approach developed by Robert Hider is to look for simple ligands which will produce a neutral ligand and a neutral iron complex and which will avoid the problems above, and being small will be easily absorbed by the gut. The ligands chosen for this study were the hydroxypyridones. 2-hydroxypyridine-N-oxide, which can be considered as an aromatic hydroxamate, has a high affinity for iron and is acid stable and this led to consideration of 3-hydroxypyridine-4-one and 3-hydroxypyridin-2-ones. As a consequence many derivatives have been synthesized and, for example, 1,2-dimethyl-3-hydroxypyridine-4-one has been used in clinical trials; it is orally active and can remove iron from overloaded patients.

2-hydroxypyridine-N-oxide   3-hydroxypyridine-4-ones

3-hydroxypyridine-2-ones

The copper complex of the simple dipeptide, $(L—His)_2$, which may be regarded as resembling the form in which copper is transported in human serum albumin, has been used successfully in the treatment of a second genetic disorder, Menke's disease. This disease is typified by rapidly progressive cerebral degeneration and spirally twisted ('kinky') hair. It is intriguing to speculate on the possibility of drugs being based on the metal sites found in the metallothioneins and phytochelatins.

7.2 The therapeutic use of preformed coordination compounds

Although preformed metal complexes have been used in medicine—gold complexes as antiarthritics, copper complexes in the treatment of rheumatism and as anti-inflammatories—it is perhaps the application of platinum complexes in cancer therapy that is the best-known example of the use of coordination complexes in the treatment of disease. Metal binding has also been said to enhance the performance of organic drugs: for example, the zinc complex of ibuprofen is more effective as an anti-inflammatory than the ligand alone and the anticancer drug bleomycin is thought to cleave DNA as an iron(II) complex via oxygen activation.

Platinum and anticancer activity

During the course of experiments designed to determine the effects of electric fields on the growth of *E.coli* bacteria, Barnett Rosenberg and his collaborators observed the unusual phenomenon of filamentous growth of the bacterial cells. Normal cell division had been inhibited and the cells grew up to 300 times their normal length.

After much study it was found that this process derived from the presence of platinum(II) and platinum(IV) ammine chloride complexes which had been generated *in situ* by electrolysis at the platinum electrodes used—the solvent solution contained ammonium chloride.

Subsequent investigations revealed that one compound responsible for filamentation was *cis*-diammineplatinum(II) chloride (*cis*-platin), a classic coordination complex, the synthesis and structure of which were well known.

Animal tests revealed that *cis*-platin was a very effective antitumour agent causing regression of both fast- and slow-moving tumours and that it also was an inhibitor of DNA (deoxyribonucleic acid) synthesis. Interestingly, the corresponding *trans*-complex has no effect against tumour growth. *Cis*-platin is now used in humans; it is not active against all cancers but is effective against testicular cancers, and active against ovarian, lung, bladder, head, neck, and cervical cancers.

The mechanism of the anticancer activity of *cis*-platin is not fully understood, but certain features have emerged. Extracellular fluids have high chloride concentrations and so replacement of the chloride in *cis*-platin by water is suppressed. Inside the cell, the chloride concentration is much diminished and so exchange of chloride by water occurs to generate products such as $[Pt(NH_3)_2(OH_2)_2]^{2+}$ and $[Pt(NH_3)_2(OH)_2]$. These can interact with the nitrogen atoms (N-7) of the guanine bases and to a lesser extent with the nitrogen atoms of the adenine bases, to produce either inter- or intrastrand *cis*-$Pt(NH_3)_2$ bridges (Fig. 7.2).

The way in which the bridges are formed between the N-7 atoms of adjacent guanines has been shown in the crystal structure of a platinum dinucleotide complex, $[Pt(NH_3)_2\{d(pGpG)\}]$ (Fig. 7.3). This type of intrastrand bridge cannot be formed by the *trans*-complex for steric reasons.

DNA consists of two strands which intertwine to give a double helix. Each strand is a condensation polymer of nucleotides containing 2'-deoxyribose as

cis-platin

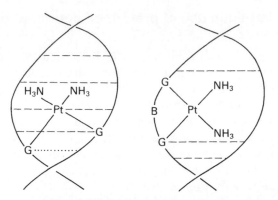

Fig. 7.2. The interaction of *cis*-Pt(NH$_3$)$_2$ fragments with DNA (reproduced with permission from Oxford University Press).

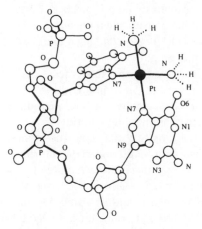

Fig.7.3. Schematic drawing of the molecular structure of *cis*-[Pt(NH$_3$)$_2${d(pGpG)}]
(reproduced with permission from Ellis Horwood).

the sugar. A nucleotide is a nucleoside in which the sugar is linked to a phosphate group and nucleosides are composed of a purine or pyrimidine base attached to the sugar ribose through the N-9 and N-1 atoms respectively (Fig. 7.4). In DNA the bases present are adenine, guanine, cytosine, and thymine and the complementary strands of DNA are stabilized by hydrogen-bonding between the base pairs.

When an organism grows, the cells divide and multiply, and the DNA doubles by a process called DNA replication. The hydrogen bonds linking the two strands of a DNA molecule break apart and each strand uses nucleotides present in the nucleus to synthesize a strand complementary to itself. The result is two daughter molecules each identical to the parent. The DNA acts as a template for the synthesis of messenger RNA in the process of transcription and this RNA carries the genetic code (Section 6.2). The interaction of *cis*-platin would interfere with these processes and it is this pertubation of the DNA structure which perhaps lies at the centre of its anticancer activity.

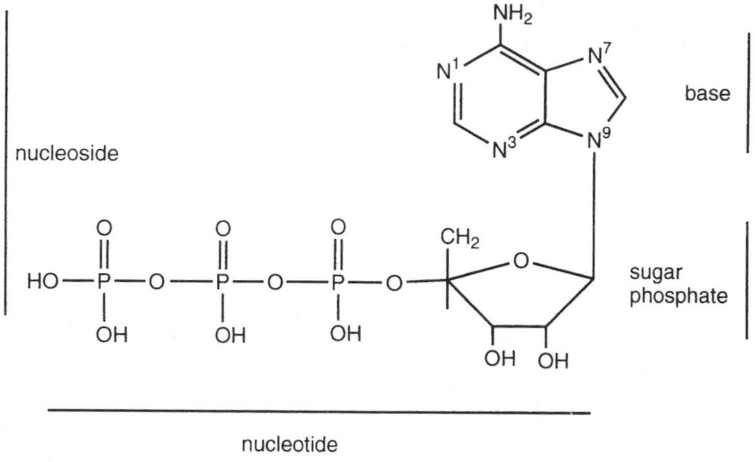

Fig. 7.4. The construction of a nucleotide; adenosine triphosphate (ATP) is used as an example.

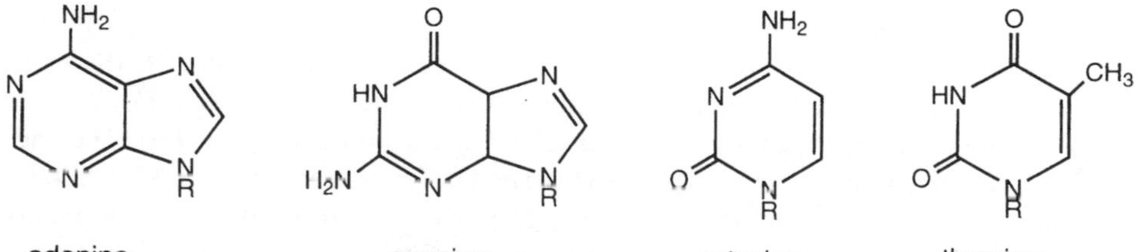

Second-generation platinum drugs have now been synthesized which are proving to produce less severe side-effects than *cis*-platin. Representative of these is diammine(1,1-cyclobutanedicarboxylato)platinum(II) (carboplatin) which is more slowly aquated than *cis*-platin.

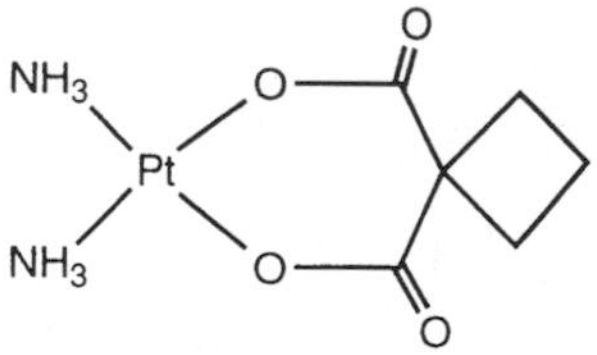

carboplatin

Gold and antiarthritic agents

Gold has always fascinated mankind and was regarded by the early civilizations as a panacea for all diseases. Modern chrysotherapy, therapy deriving from gold, was pioneered by the French physician Forestier who, in 1929, first introduced gold(I) thiolates as therapeutic agents for rheumatoid arthritis. Today all the gold complexes used in medicine are derived from

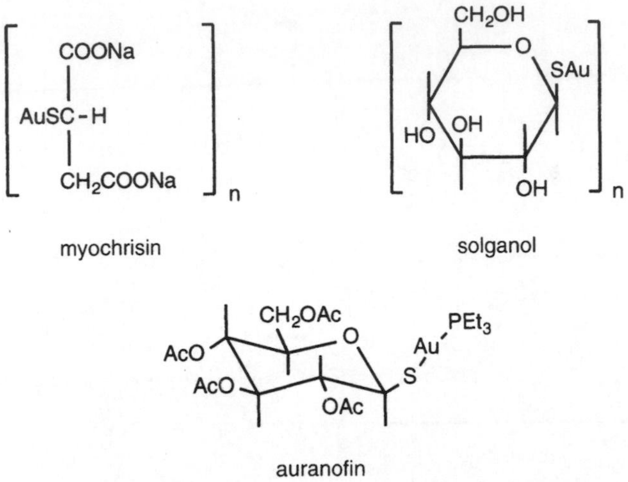

myochrisin

solganol

auranofin

gold(I). Representative gold(I) antiarthritic drugs are myochrosin [sodium aurothiomalate] and solganol (aurothioglucose], which are injected intramuscularly, and auranofin [(*S*)—2,3,4,5-tetraacetyl-1-D-thioglucose-(triethylphosphine)gold(I)], which is administered orally.

The 1:1 gold thiolates have not yet been crystallised but NMR, mass spectrometric, and X-ray scattering studies have revealed that they are not monomeric but are built up of ring and chain structures (Fig. 7.5). They are administered in doses of *c*.25mg per week for several years and it can take several months before beneficial effects show. Consequently several grams of gold can be used in the treatment and can remain in the body long after the therapy is completed.

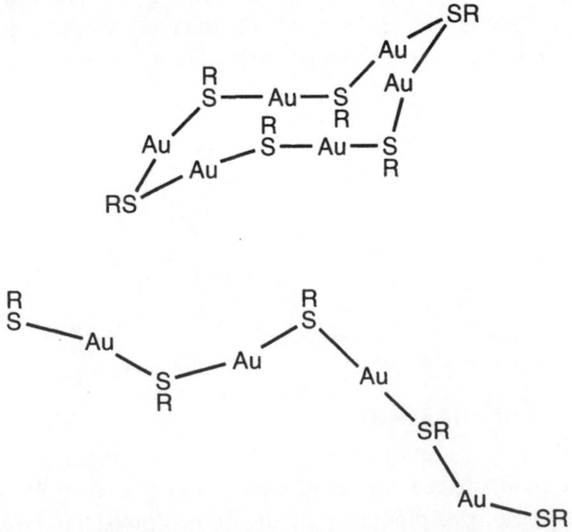

Fig. 7.5. Ring (top) and chain (bottom) structures of gold(I) thiolate drugs.

In contrast auranofin is a lipid-soluble, linear, two-coordinate complex. It may be hydrolysed during absorption into the intestine so that the species which then enters the blood stream is $Au(I)(PEt_3)(\beta\text{-D-thioglucose})$.

The end products of the metabolism are likely to be the same as for the thiolates. The precise modes of action of these compounds are uncertain but it is thought that antienzyme and anti-inflammatory activities are involved.

Gold(III) is isoelectronic with Pt(II) and has isostructural complexes; it is therefore appropriate to ask why Au(III) is not used as an anticancer agent. The simple answer is that *cis*-diamminedihalogold(III) analogues of *cis*-platin have not yet been characterized; reaction of $[Au(NH_3)_3X]^{2+}$ and $[Au(NH_3)_4]^{3+}$ with halide anions (X^-) give only the *trans*-product. Furthermore ligand substitutions occur much more readily in Au(III) complexes, which are also stronger oxidants than Pt(II) complexes. Aqua complexes of Au(III) such as $[AuCl_3(H_2O)]$ are highly acidic, and simple anions such as $[AuCl_4]^-$ can oxidise, for example, methionine to the sulphoxide and cystine disulphides to sulphonates and so cannot be used as drugs.

Endpiece

Biocoordination chemistry is a living subject and so, like life itself, is full of surprises. This has been demonstrated in the text by the many unusual, and often unexpected, coordination environments found at metallobiosites. Reinforcement of this expression came during the proof stage of this text when the crystal structure of the nickel-iron hydrogenase (6.7) from *Desulphovibrio gigas* was announced. This showed the presence of two $4Fe-4S^*$ clusters and one $3Fe-4S^*$ cluster, arranged as 'stepping stones' for electron transfer, together with a hydrogen-binding site which, quite unexpectedly, appears to be a heterodinuclear iron–nickel cluster. The nickel is bound to four protein cysteine ligands and the iron to two bridging cysteines together with three or four as yet unidentified non-protein ligands. Nature continues to confound the chemist and by doing so provides us with many challenging problems.

Further reading

Any book such as this can only provide a brief insight into an area. Below is a list of sources which will help the reader with any further research.

Textbooks
Most contemporary textbooks on inorganic chemistry contain a chapter on aspects of bioinorganic chemistry. More comprehensive texts are listed below.

— J.J.R. Fraústo da Silva and R.J.P. Williams, *The Biological Chemistry of the Elements*, Oxford University Press, 1991.

— R.W. Hay, *Bioinorganic Chemistry*, Ellis Horwood, Chichester, 1984.

— M.N. Hughes and R.K. Poole, *Metals and Micro-organisms*, Chapman & Hall, London, 1989.

— K. Burger (ed.), *Biocoordination Chemistry — Coordination Equilibria in Biologically Active Systems*, Ellis Horwood, Chichester, 1990.

— S.J. Lippard and J.M. Berg, *Principles of Bioinorganic Chemistry*, University Science Books, Mill Valley, California, 1994.

— W. Kaim and B. Schwederski, *Bioinorganic Chemistry: Inorganic Elements in the Chemistry of Life,* John Wiley & Sons, Chichester, 1994.

— I. Bertini, H.B. Gray, S.J. Lippard and J.S. Valentine, *Bioinorganic Chemistry*, University Science Books, Mill Valley, California, 1994.

There are also many monographs, such as the series *Metal Ions in Biology*, ed. T.G. Spiro, Wiley, New York, which are dedicated to the bioinorganic chemistry of individual metals or subject areas.

Review articles
In addition to textbooks, there is also available an abundant review literature concerning the topics discussed in this book. These take the reader deeper into the subject area. Dedicated series include the two below.

— *Metal Ions in Biological Systems*, ed. H. Sigel, Dekker, New York.

— *Advances in Inorganic Biochemistry*, ed. G.L. Eichhorn and L.G. Marzilli, Elsevier, New York.

Individual review articles can be found sometimes in special editions of series such as those given below.

— *Progress in Inorganic Chemistry*, ed. K. D. Karlin, Wiley.

— *Advances in Inorganic Chemistry*, ed. A. G. Sykes, Academic Press.

— *Structure and Bonding*, Springer-Verlag, Berlin.

Journals such as *Angewandte Chemie*, the *Journal of Chemical Education, Coordination Chemistry Reviews, Accounts of Chemical Research, Chemical Reviews* and *Chemical Society Reviews* are also excellent sources of review articles and the *Journal of Chemical Education* published a collection of reviews entitled 'Bioinorganic Chemistry - State of the Art' in 1985 (**62**, 916-1001). There is also an extremely readable section entitled 'Biological and Medicinal Aspects' (of coordination chemistry) in **6** of *Comprehensive Coordination Chemistry*, ed. G. Wilkinson, R. D. Gillard and J. A. McCleverty, Pergamon Press, 1987.

As well as the chemistry-based literature, there is also the biology-based literature Excellent short, state-of-the-art, reviews can be found regularly in *Trends in Biological Sciences.* and more specialist subject area reviews are to be found in sources such as *Advances in Protein Chemistry, Annual Reports of Biochemistry*, and *Annual Reviews of Biophysics and Biophysical Structure.*

Figure acknowledgements

2.1 Heald, S.M. *et al.* (1979). *J. Am. Chem. Soc.*, **101**, 67. © 1979 American Chemical Society.

2.2 Rice, D.W. *et al.* (1983). Adv. Inorg. Biochem., **5**, 39.

2.3 Lippard, S.J. (1986). *Chem. in Brit*, 1986, 227

2.4 Baker, E.N. *et al.* (1990). *Pure Appl. Chem.*, **62**, 1067.

2.5 Hughes, M.N. (1987). Chapter 62.1 in *Comprehensive coordination chemistry* ed. G. Wilkinson. Pergamon Press, Oxford. *© 1987 Pergamon Books Ltd.*

2.7 Karpistin, T.P. and Raymond, K.N. (1992). *Angew. Chem. Int Ed. Eng.*, **31**, 467.

2.11 Armstrong, E.M. *et al.* (1993) *J. Am. Chem. Soc.*, **115**, 807.

3.5 Shanaan, B. (1982). *Nature*, **296**, 683.

3.6 Chem. Eng. News (1985), 45. *© 1985 American Chemical Society.*

3.7 Wilkins, R.G. (1992). *Chem. Soc. Revs.*, 173.

4.1 Norris, G.E. *et al.* (1986) *J. Am. Chem. Soc.*, **108**, 2784.

4.3 Karlin, K.D. *et al.* (1989). Inorg. Chim. Acta, **165**, 37.

4.6 Beinert, H. and Kennedy, M.C. (1989). *Eur. J. Biochem.*, **186**, 5.

4.9 Dickerson, R.E.

4.13 Huheey, J.E. *et al.* (eds) (1993). Inorganic chemistry: principles of structure and reactivity, 4th edn. *Copyright © 1993 Harper Collins College Publishers.*

4.14 Klein, M. *et al.* (1991). *J. Inorg. Biochem.*, **43**, 363.

5.1 Fontecave, M. and Pierre, J-L. (1993). *Bull. Chim. Soc. Fr.*, **130**, 77.

5.3 Hughes, M.N. (1987). Chapter 62.1 in *Comprehensive coordination chemistry* ed. G. Wilkinson, *et al.* Pergamon Press, Oxford. *© 1987 Pergamon Books Ltd.*

5.9 Koch, S. *et al.* (1975). *J. Am. Chem. Soc.* **97**, 916.

5.10 Solomon, E.I. *et al.* (1994). *Chem. Rev.*, **94**, 827.

5.11 Rosenzweig, A.C. *et al.* (1993). *Nature*, **366**, 537.

5.14 Que, L. Jr. and True, A.E. (1990). *Progr. Inorg. Chem.*, **38**, 102 *© 1990 John Wiley & Sons Inc.*

6.4 Christianson, D.W. and Lipscomb, W.N. (1989). *Accts. Chem. Res.*, **22**, 68.

6.5 Vallee, B.L. *et al.* (1991) *Proc. Nat. Acad. Sci. USA.*, **88**, 999.

6.16 Sellman, D. (1993). *Angew. Chem. Int. Edn. Engl.*, **32**, 64.

6.18 Coucouvanis, D. (1991). *Accts. Chem. Res.*, **24**, 1.

6.19 Chan, M.K. *et al.* (1993). *Science*, **260**, 792. *© 1993 American Association for the Advancement of Science.*

7.3 Kendrick, M.J. *et al. Metals in biological systems*, Ellis Horward Ltd., Chichester. *© 1992. Ellis Horward Limited.*

4.4, 4.5, 4.10, 4.11, 4.15, 6.24 and 7.2 Fraústo da Silva, J.J.R. and Williams, R.J.P. (1991). *The biological chemistry of the elements*. Oxford University Press.

Formulae 4.3, 4.5, 4.6 (p.44) Weighardt, K. (1989). *Angew. Chem. Int Ed. Eng.*, **28**, 1168.

Index